T0215028

Whitestein Series in Software Agent Technologies

Series Editors:
Marius Walliser
Stefan Brantschen
Monique Calisti
Thomas Hempfling

This series reports new developments in agent-based software technologies and agent-oriented software engineering methodologies, with particular emphasis on applications in various scientific and industrial areas. It includes research level monographs, polished notes arising from research and industrial projects, outstanding PhD theses, and proceedings of focused meetings and conferences. The series aims at promoting advanced research as well as at facilitating know-how transfer to industrial use.

About Whitestein Technologies

Whitestein Technologies AG was founded in 1999 with the mission to become a leading provider of advanced software agent technologies, products, solutions, and services for various applications and industries. Whitestein Technologies strongly believes that software agent technologies, in combination with other leading-edge technologies like web services and mobile wireless computing, will enable attractive opportunities for the design and the implementation of a new generation of distributed information systems and network infrastructures.

www.whitestein.com

Javier Vázquez-Salceda

The Role of Norms and Electronic Institutions in Multi-Agent Systems

The HARMON*IA* Framework

Birkhäuser Verlag
Basel · Boston · Berlin

Author's address:

Javier Vázquez-Salceda
Intelligent Systems Group
Institute of Information and Computing Sciences
Utrecht University
P.O. Box 80.089
3508 TB Utrecht
The Netherlands
e-mail: javier@cs.uu.nl

2000 Mathematical Subject Classification 68T05, 68T27, 68M14

A CIP catalogue record for this book is available from the
Library of Congress, Washington D.C., USA

Bibliographic information published by Die Deutsche Bibliothek
Die Deutsche Bibliothek lists this publication in the Deutsche Nationalbibliografie;
detailed bibliographic data is available in the Internet at <http://dnb.ddb.de>.

ISBN 3-7643-7057-2 Birkhäuser Verlag, Basel – Boston – Berlin

© 2004 Birkhäuser Verlag, P.O. Box 133, CH-4010 Basel, Switzerland
Part of Springer Science+Business Media
Cover design: Micha Lotrovsky, CH-4106 Therwil, Switzerland
Printed on acid-free paper produced from chlorine-free pulp. TCF∞
Printed in Germany
ISBN 3-7643-7057-2

9 8 7 6 5 4 3 2 1   www.birkhauser.ch

Contents

    6.6   Summary 205

7   Conclusions   207
    7.1   Our Proposal of a New Framework 208
    7.2   Original Contributions 211
        7.2.1   Distinction between Normative and Operational 211
        7.2.2   Normative Systems and Contexts 211
        7.2.3   New Terminology 212
        7.2.4   Connection between Formal Specification and Agent Implemen-
                tation 212
        7.2.5   Norm Enforcement as Detecting Illegal Worlds 213
    7.3   Ongoing Work 213
        7.3.1   Definition of a Modular Architecture for E-Organizations . . . 213
        7.3.2   Creation of Tools for E-Organizations 217
        7.3.3   Testing the Framework in New Domains 219
    7.4   Suggestions for Further Research 219

III   Appendix and Bibliography   223

A   Medical Data Protection and the Internet   225
    A.1   Medicine, Information Technology and Privacy 225
    A.2   Advantages of Electronic Formats for Medical Data 227
    A.3   Requirements to be Fulfilled 227
        A.3.1   Requirements from the Medical Community 228
        A.3.2   Law-Enforced Requirements 228
    A.4   Privacy and Security of Electronic Medical Data 232
    A.5   Security Measures for Medical Information Systems 233
    A.6   Desirable Characteristics of a Medical Information System 234
    A.7   The New Users of Medical Data 236
    A.8   Summary 236

B   The UCTx System   239
    B.1   Creating a Multi-Agent System for a Hospital's Transplant Unit . . . 239
    B.2   An Example: The Cornea Transplantation 244
    B.3   Summary 248

    Bibliography 250
    List of Acronyms 269
    Glossary 271
    Authors Index 273

List of Figures

List of Tables

Preface

In recent years, several researchers have argued that the design of multi-agent systems (MAS) in complex, open environments can benefit from social abstractions in order to cope with problems in coordination, cooperation and trust among agents, problems which are also present in human societies.

The *agent-mediated electronic institutions* (*e-institutions* for short) is a new and promising field which focuses in the concepts of norms and institutions in order to provide normative frameworks to *restrict* or *guide* the behaviour of (software) agents. The main idea is that the interactions among a group of (software) agents are ruled by a set of explicit norms expressed in a computational language representation that agents can interpret. Such norms should not be considered as a negative constraining factor but as an aid that guides the agents' choices and reduces the complexity of the environment making the behaviour of other agents more predictable.

However, there are several issues to be solved. The main one is that current work is either too *theoretical*, focused on norm formalization by means of very expressive logics that are computationally hard, or too *practical*, focused on the implementation of *e*-institutions but losing accuracy and expressiveness in the normative system. No connection between both approaches has been defined. Another issue that researchers in the field face is the variety of terminologies (i.e., different terms for the same concept, or the same term with different meaning) that are in use. Finally, an additional issue is that, despite the fact that a significant amount of theoretical and practical work has been carried out in *e-institutions'* definition, formalization and implementation, they have been applied to small or quite simple experimental setups.

The contents of this book are an extract and extension of the author's PhD thesis [213]. The aims of the thesis were twofold: i) to define a unifying framework where the formalization of the normative system in terms of logics and the final implementation of the multi-agent system are connected, filling the existing gap between both approaches, and ii) to test the possibility to formalize an electronic institution in complex, highly regulated domains such as international *e*-business or medicine. As case study, we chose the allocation of human organs and tissues for transplantation purposes. Such real setup not only has to cope with the (typical) problems that arise in medical applications (because of the inherent complexity of medical knowledge engineering) but also should include the variety of regulations (national and international) that apply to organ and tissue allocation/transplantation

Approach and Contributions

Our main observation is that, in human legal systems, norms are specified in regulations that are (on purpose) at a high abstraction level. The level of abstraction is high in order not to be dependent on a circumstantial implementation of the norm. Norms should be stable for many situations and for a relative long time. Therefore, it is obvious that the norms do not have concrete handles for their direct implementation on MAS. In order for the norms to be implemented, these norms should be translated to more concrete, operational representations, that can be used by agents in their decision making processes.

The main novelty of this work is the approach that was taken to create the new framework. Instead of trying to define a single language to model all the normative aspects of a given domain, we propose a multi-level framework, filling the gap between theoretical and practical approaches, by:

- identifying the different abstraction levels involved in a normative framework (values, norms, rules, procedures),

- defining an appropriate language for each level,

- defining a connection between these levels, and

- identifying the influence of context for each abstraction level.

We argue that such a connection among levels is beneficial both from the individual agents' point of view (the *micro-level*) and from the multi-agent systems' view (the *macro-level*) :

- in the case of agents, the connection between abstraction levels within the normative framework allows that the agents entering into an *e*-institution use either the defined protocols, the rules, the norms or a combination of them to guide their behaviour;

- in the case of multi-agent systems, the connection between abstraction levels fills the gap between formal specification of norms and the final implementation, with effects both in the design and the verification of the resulting implementation.

The main contributions of the work presented in this book can be summarized as follows:

- We identify the different abstraction levels involved in a normative framework and define a connection between these levels. The connection between levels allows to connect formal specification of norms with the final implementation, filling the existing gap between theoretic and practical approaches. Also the influence of context for each abstraction level is analysed.

- We present HARMON*IA*, a framework that includes all the identified levels, and which allows us to study the effects from the individual agents and the whole multi-agent system perspective.

  - In the case of agents, we present a special kind of *Deliberative Normative Agent*, the *Flexible Normative Agent*, which use the connection between norms, rules and protocols in order to follow the defined protocols (which is time-efficient) but being able to regularly check the related rules and norms in order to cope with situations that were not foreseen by the protocol and to recover from them.

  - In the case of multi-agent systems, the connection between abstraction levels allows to enforce norms by means of the related violations and sanctions, expressed in terms of conditions that can be checked by the multi-agent system. We show how norm enforcement can be done by means of a special kind of agent, the *Police Agent*.

- We propose an accurate terminology for the concepts involved in the e-institution field, including terms such as norm, rule, procedure, protocol, policy and context. Such terminology is the result of a conceptualization made by studying the role of norms and their effects from the agent's and the society's perspective.

Additional Reading

In this book we combine three different Artificial Intelligence disciplines (Software Agents, Artificial Intelligence and Law, Artificial Intelligence in Medicine) to propose our framework. Although we will do some review on existing approaches, we will be highly selective, only mentioning those that are closer to the needs of the objectives of this work.

Software Agents is a wide area in Artificial Intelligence, and there are too many approaches, systems and applications to be reviewed in a single book. [131] is not an in-depth review, but presents a roadmap on the current topics of research in agent technology and some technological challenges to be addressed in the future. The reader interested on a more introductory material can read the book by Wooldridge [223].

Artificial Intelligence and Law (usually referred to as AI & Law) is an interdisciplinary field with a wide research community that applies AI methods for representing jurisprudential theories and/or to build AI-based information systems to assist legal decision makers in their practice. On Chapter 3 we present a review on some approaches from AI & Law that can be applied to agent technologies. The reader who wants a broader view on AI & Law can find an interesting overview in [172]. [106] is a classic article on formalization of rule-based legal systems, while [7] discusses the issues of reasoning in precedence-based legal systems. In-depth discussions on legal reasoning (including related issues such as defeasibility) can be found in [169] and [95]. Finally [87] is a good book on argumentation.

Artificial Intelligence and Medicine (usually referred to as AI & Medicine) is another wide area of research, as it presents lots of challenges that are the perfect test-bed for a wide range of AI methods and approaches. In Chapter 2 we present a small review of some representative examples on the use of agents in Medicine. As far as we know there is no single book that introduces or reviews all the different (AI-based clinical decision-making; medical knowledge engineering; intelligent medical information systems, databases and libraries; intelligent devices and instruments). Elsevier Science publishes a journal on all these aspects, and it is available on-line [6]. Fox and Das discuss in [75] the safety and soundness requirements of any application of Artificial Intelligence to the medical domain. Finally [146] is a good compilation on some applications of agents to health care.

Organization

This book tries not only to present the final results of the research (the HARMON*IA* framework) but also to transmit to the reader some details about the experiences of the author during the process. After an overview of the problem, the book presents the first attempt to design and implement a system by using an existing formalism. Then the problems that arose are presented and the need for another approach is introduced. The problem is then analysed from several points of view: a) norm theories, to study the view several branches of science (Legal Theory, Sociology, Economics and, of course, Computer Science) have about Norms; and b) the use of norms in Multi-Agent Systems. Then a new framework for Agent-Mediated Electronic Institutions is introduced, aiming to cope with the problems previously identified.

This book is structured in seven chapters (including the introduction one), and two appendixes.

Chapter 1 is a brief introduction to the *e*-institutions area. First we introduce the reader to the problem of control *versus* autonomy, specially arising in open multi-agent systems. Then we present the concepts of *coordination*, *coordination theories*, *social structures*, *commitments* and *conventions*, to end with the definitions of *human* and *electronic institutions*.

Chapter 2 presents our first attempt on applying electronic institutions to a complex domain such as organ and tissue allocation for transplantation purposes. First we introduce the problem of organ and tissue allocation, and we review some of the agent technologies that have been applied to Medicine. Then the *CARREL* system –a multi-agent system to mediate in organ and tissue allocation– is presented. The specification process is explained by using ISLANDER [68], a language to formally define agent-mediated electronic institutions. We will show the resulting formal description, the limitations that arise with such a language, and the need for more complex, higher level formalisms.

Chapter 3 reviews the different interpretations of the concept of norms that can be found in different research fields, from Sociology and Socioeconomics to Legal Theory. Then we present some of the research focused on norm specification and modelling, agent theories with norms, normative agents' architectures and electronic institutions.

Chapter 4 introduces the HARMON*IA* framework. The chapter focuses on the study of norms from the agent point of view. The relation among the norms and the agent's beliefs, knowledge and goals is described. By doing so we will identify some of the elements that compose our framework (*norms, rules, procedures, policies, roles, context*) and we will provide an accurate definition. Also the effects of norms in the standard BDI cycle are analysed. Finally we will see which are the attributes that *Police Agents* should have to enforce norms.

Chapter 5 extends the HARMON*IA* framework by the description of the different levels of abstraction that compose a normative framework from the organizational perspective. We describe the process to build an agent-mediated normative electronic organization, from the statutes of the organization to the norms to the final implementation of those norms in rules and procedures. In this chapter the proposed terminology is completed.

Chapter 6 presents a quite extensive case study of the HARMON*IA* framework in the context of human organ and tissue allocation. We will analyse which are the values, norms, rules and procedures to be followed by the agents and then we will be able to validate and refine the *CARREL* electronic institution presented in Chapter 2.

Chapter 7 provides a summary of the work in the thesis and its contributions, presents a small report on our ongoing work to extend the framework and then proposes some of the lines for future work.

Appendix A describes which are the requirements (both from clinician and from Spanish and European Law) to be fulfilled by medical applications in Spain, which are (with minor adjustments) valid in the rest of the European Union.

Appendix B describes the *UCTx*, an agency that acts as a mediator between hospitals and the *CARREL* electronic organization presented in Chapter 2 and Chapter 5.

Acknowledgements

To Ulises Cortés, who has been advisor, mentor and friend. Lots of the ideas presented in this book were inspired on our continuous discussions about the nature of norms and its effects on agents. In the very spirit of this work, while I was doing my PhD he sought to give me *autonomy* enough to explore new paths while providing some *control* to keep me from losing the sight of the track. And he succeeded. No words are enough to thank you.

Very special thanks to Julian Padget and Frank Dignum. This work would never been possible without their valuable contribution and enthusiasm. Thanks also to the people in the Computer Science Department at University of Bath, who hosted me during my visit in Bath, and improved my communication skills in English. I would also like to thank all the people I met in my first visit at the Institute of Information and Computing Sciences at Utrecht University, in particular to all those who let me participate in their ongoing discussions about organizational and normative modelling: John-Jules Meyer, Virginia Dignum, Davide Grossi, Huib Aldewereld and Ludger van Elst. Some of the ideas that are presented in this work were born or highly sharpened during those discussions.

I would like to thank Sabrine Lopes, Francisco Vázquez, Mario Nicolás and David Busquets, who took part in the first brainstormings that gave birth to the *CARREL* idea. I would also like to thank all the students that have been involved in the development of the *CARREL* project: Jordi Anglí, Jorge Castillo, Corentin Massot, Damien Boissou, Joan Cervera, Marc Valerio, Joan Torras and David Cabanillas. They are the ones that had to cope with my *weird* ideas and made them become *real*. Thanks also to Owen Cliffe, Daniel Jiménez Pastor and Albert Miguel Turu, from the Agents Group in Bath, which are working on the implementation of the HARMON*IA* platform.

Thanks to the staff at the Banc de Teixits, Hospital de la Santa Creu y Sant Pau,and specially to Dr. Antonio López-Navidad and Dr. Francisco Caballero, who shared with us a bit of their wisdom to give us a better understanding on organ and tissue transplantation. Thanks also to the staff members of the Organització Catalana de Trasplantaments, and specially to Dr. Jordi Vilardell, Marga Sanromá, Montse Clèries and Luz Amado, who allowed us to interrupt their daily coordination work and gave us precise details on the complex process of organ and tissue allocation.

Thanks also to John Fox, Liz Black and the rest of the researchers at Cancer Research UK. The work we have done together has been very valuable in improving the soundness of the *CARREL* specification, and hopefully it will be only the beginning of more collaborations in the future.

To all my former colleagues in the KEMLg group in Barcelona and my new colleagues at the Intelligent Systems group in Utrecht, for all the work (and the fun) we shared, and for their support and friendship. Special thanks to Steve Willmott, who found time in his busy schedule to check preliminary versions of the work, coming back with a full list of comments and annotations. Thanks too to Henry Prakken for his comments on current research on AI & Law.

Thanks to all the (anonymous) reviewers that have contributed with their comments in different stages of this work and on preliminary versions of this document.

To my friends, to make me feel so fortunate and remind to this workaholic that *"we should work to live, not live to work"*.

Last, but certainly not least, thanks to my family. To my parents, who have provided anything they could to allow me explore my goals and dreams. To Ivo, Mari, Julia, Francisco, Elena y David, for all their love and their unconditional support. Os quiero.

This work has been partially funded by the European Commission through the IST-1999-10176 A-TEAM project and the IST-2000-28385 Agentcities.NET project.

Part I

Applying Electronic Institutions and Norms to Complex Domains

Chapter 1

Introduction

In recent years, multi-agent systems have not only increased in number, spreading through several domains such as Electronic Commerce, Health Care or resource allocation, but the MAS approach is even becoming a promising new programming paradigm called *Agent-Oriented Software Engineering* (AOSE) This field is based in the concept of *software agent*. Wooldridge and Jennings [226] define an *Agent*[1] as a computer program capable of taking its own decisions with no external control (*autonomy*), based on its perceptions of the environment and the objectives it aims to satisfy. An agent may take actions in response to changes in the environment (*reactivity*) and also it may take initiatives (*proactivity*). A further attribute of agents is their ability to communicate with other agents (*social ability*), not only to share information but, more important, to coordinate actions in order to achieve goals for which agents do not have plans they can fulfill on their own, solving even more complex problems[2]. All these attributes suggest that multi-agent systems are well-suited for solving complex problems where coordination among several entities might be valuable in order to find an optimal or near to optimal solution with limited time and computational resources. Agents can also exhibit other useful attributes such as *learning, rationality, mobility, benevolence* and *veracity*. A group of agents that interact among themselves and share some common objectives is called *agency*.

The capacity of Multi-Agent Systems (MAS) to distribute the work among several autonomous, rational and proactive entities (the software agents) and even to adapt and re-organize themselves in some extent, is one of the reasons that has lead this technology to be considered a *leading* option for the implementation of open distributed systems in complex domains such as electronic commerce or organizational modelling (computational models that try to emulate a given organization). But building multi-agent systems is a very complex process. The simplest scenarios are the *closed multi-agent systems* (also known as *self-contained multi-agent systems*). The designer(s) of such system aim(s) to

---

[1]From now on we will use the term *agent* to talk about *software agents*, unless another meaning is noted explicitly in a particular context.

[2]Such communities of agents that show social abilities are sometimes called *Intelligent Agent Societies* (IAS).

establish a closed society of autonomous agents that interact among themselves in order to achieve one or several goals, as none of them alone has all the capabilities needed to solve the entire problem at hand. In this scenario the multi-agent system is usually built by a single developer or a single team of developers, and the chosen option to reduce complexity is to ensure cooperation among the agents they build including it as an important system requirement. In this approach, the designing process of what can be tagged as a *cooperative problem solving system* is driven by the definition of a minimum set of common goals which are called *system goals* or *global goals*. Such goals are the way to guide the systems' behaviour towards the desired solution. Developers focus their efforts to ensure that all agents|entities in the system, although they may have different goals, agree on the global goals, that is, that all the agents include in some way those global goals inside their individual goals. In fact the advantages of this scenario are not only reduced to goal agreement: having a single developer or a team of developers also makes it easier (i) to solve communication issues (by defining an ontology shared by all agents) and also (ii) to ensure the full coverage[3] of all the tasks by the MAS. However, it is not always easy to find a way to model the system as a set of individual goals, and ensure that the individual behaviour of each agent and the aggregation of such behaviours resulting from agents' interaction will lead to the (desired) global behaviour.

This problem becomes more challenging in *open multi-agent systems*, where there are several, heterogeneous agents created by different people. In such scenarios agents may not only have completely different goals but there is also a higher probability that they have conflicting goals, that is, the individual goals of an agent interfering or competing with other agents' goals. Another related issue is *trust*. In [78] Gambetta defines *trust* as a particular level of subjective probability with which an agent a_j will perform a particular action before the action is performed. In the previous scenario, where cooperation among agents is included as part of the designing process, there is an implicit trust: an agent a_i requesting information or a certain service from agent a_j can be sure that such agent will answer him if a_j has the capabilities and the resources needed, otherwise a_j will inform a_i that it cannot perform the action requested. However, in an open environment *trust* is not easy to achieve[4], as agents may give incomplete or false information to other agents or betray them if such actions allow them to fulfill their individual goals. In such scenarios developers use to create *competitive systems* where each agent seeks to maximize its own expected utility at the expense of other agents.

In summary, the creation of open distributed systems has to cope with several issues:

- Heterogeneity among members: open systems allow to each of the participants to develop their own agents. Although this option is very positive from the participants' point of view, it becomes a serious problem from the system's point of view, while defining and building the communication mechanisms and protocols, as each

---

[3]The *Coverage* of a MAS relative to a task reflects the amount of the task that can be performed by the set of agents.

[4]There are lots of work to try to define reputation and trust mechanisms in open systems, such as [1, 229, 231, 189].

participant has its own model of the world, its own reasoning and decision making mechanism.

- Communication network issues: the use of public communication channels such as *Internet* makes extremely easy to provide connectivity to people and computer systems, *reducing the distance* among them. But the savings in infrastructure that come with the use of existing networks leads to soundness and security issues related to the architecture of those networks.

- Participants' trust: in this scenario the implicit trust and cooperation of previous scenario does not exists: agents may *betray* other agents by accepting to perform an action and then deciding to do something else (even the opposite action) if doing so better fits its own goals. The lack of trust comes mainly from i) the *heterogeneity* of the agents in the system, and ii) the high level of *anonymity* caused by having all the interactions among the participants made through distance

A central problem that a MAS developer should face is the following: How to ensure an efficient and acceptable behaviour of the system without diminishing the agents' *autonomy*. So we are facing a compromise between *Control* (of the system) and *Autonomy* (of the agents).

1.1 Autonomy

Before introducing and analysing the concept of *coordination* and its central role in controlling agent's behaviour, let us first define *autonomy*. In Wooldridge and Jennings' definition of *Agent* [226] they state that the main characteristic of agents is that they are *autonomous*, and *autonomy* is defined as the capability of an agent to act independently, exhibiting control over its own internal state.

There are other definitions of *autonomy* in MAS. A good review is presented by Verhagen in [216]. In that work he defines *autonomy* as the degree in which one agent is independent with respect to another entity, be it the environment, other agents or even its own developer. This definition is complementary to the one by Wooldridge and Jennings, as it lists some of the sources of external control which may reduce the autonomy of a given agent. Verhagen also notes that autonomy should be considered both at the *individual level* and at the *MAS level*. At the level of individual agents the following types of autonomy are identified:

- *Autonomy from stimuli*: the degree to which the behaviour of an agent is directly controlled by external *stimuli* from other entities at runtime.

- *Cognitive autonomy*: the degree to which the agents' choices (and the actions based in such choices) are governed by other agents.

- *Learning autonomy*[5]: the degree to which agents are capable to learn new rules of behaviour, making the agent to diverge from predefined decision making rules.

---

[5]Verhagen refers to this kind of autonomy as *norm autonomy* at the individual level.

At the level of the whole Multi-Agent System, Verhagen identifies at least two types of autonomy:

- *Norm autonomy*: the degree to which the agent's choices (and the actions based on these choices) are governed by social influences.

- *Designer autonomy*: it refers to the autonomy of the agents forming a MAS with respect to the designer(s). Such autonomy is related to the self-organizing capabilities of a multi-agent system, which may lead to *emergent* behaviours.

Even though these definitions show different kinds of autonomy, there are not clear divisions among them, as some of the definitions overlap with others (for instance the definition of *norm autonomy* slightly varies from the one of *cognitive autonomy*). Therefore an accurate definition of autonomy hierarchy is needed. After studying several proposals of autonomy hierarchies [31, 45, 54, 222], Verhagen proposes four levels of autonomy:

- *Reactivity*: at this level perception is directly coupled to action. There is no autonomy at the individual level, and the only autonomy available is the one the MAS as a whole may have. Agents with this level of autonomy are called *reactive agents* [45, 222] .

- *Plan autonomy*: at this level the agent's activities are always oriented by an external imposed goal that cannot be changed by the agent itself. The autonomy of these agents (called *plan autonomous agents* [45]) is reduced to the choice of actions to achieve the goal (by some kind of means-ends reasoning).

- *Goal autonomy*: at this level the agent's goals are not just a result of the goals imposed by its context (e.g., goals coming from other entities' requests) but a composition of its own goals plus the ones coming from the context. Such agents (called *goal autonomous agents* [45]) have the autonomy to determine preferences among goals.

- *Norm autonomy*: at this level agents choose which goals are legitimate to pursue, based on a given set of norms. Such agents (called *norm autonomous agents*[6] or *deliberative normative agents* [36, 21]) may judge the legitimacy of its own and other agents' goals. Verhagen defines autonomy at this level as the agent's capability to change its norm system when a goal conflict arises, thereby changing priorities of goals, abandoning a goal, generating another goal, etc. F. Dignum, in [55], provides another view of autonomy at the norm level, allowing the agents to violate a norm in order to adhere to a private goal that they consider to be more profitable, including in such consideration the negative value of the repercussions such violation may have[7].

In most of the work in autonomous agents the degree of autonomy of an agent is fixed, that is, the agent has the same level of autonomy during all its existence. An interesting alternative is the concept of *Adjustable Autonomy* [110, 64, 217] : autonomous

---

[6]As Verhagen notes, the *norm autonomous agents* are comparable to Dennet's moral personhood [54].

[7]We will talk more about the *Norm Autonomous Agents* in Section 3.2.2.

MAS which adapt the level of autonomy to the situation at hand, depending on attributes such as the complexity of the problems to be solved or the dynamics of the MAS (the time of autonomous operation, the number of autonomous subsystems in the MAS, etc.).

To close the discussion about autonomy it is important to note that, even in the highest level of autonomy (the *norm autonomy*), the agents in a Multi-Agent system cannot be said to be completely autonomous, as they depend in more or less degree on other agents to solve a given problem[8]. In this sense Castelfranchi states in [34] that an agent's autonomy is necessarily limited, since it is situated.

1.2 Coordination and Coordination Theories

Coordination could be defined as the process of managing dependencies between activities. By such process an agent reasons about its local actions and the foreseen actions that other agents may perform, with the aim to make the community to behave in a coherent manner. Coordination consists of a set of mechanisms necessary for the effective operation of a MAS in order to get a well-balanced division of labour (*task allocation techniques*) while reducing logical coupling and resource dependencies of agents.

Lots of empirical and theoretical work has been and is currently being done to study coordination, not only for specific domains but in a more generic, domain-independent view [9]. Some of this work lead to the creation of coordination theories. A *Coordination Theory* can be defined as a set of axioms and the analytical techniques used to create a model of dependency management. Examples of coordination theories are *joint-intentions* theory [41], theories about shared plans [91] or domain-independent teamwork models [203, 187, 105, 228].

One way to tame the complexity of building a MAS is to create a centralized controller, that is, a specific agent that ensures coordination. *Coordinator agents* are agents which have some kind of control on other agents' goals or, at least, on part of the work assigned to an agent, according to the knowledge about the capabilities of each agent that is under the *Coordinator Agent*'s command. From the developer's point of view, with this approach complexity is reduced as the ultimate goal of the system is ensured by the goals of the coordinator, which supersedes the goals of the other agents in the system. Even though these kind of multi-agent architectures are easier to build, the main disadvantages of this approach come from its *centralized control*:

- the *Coordinator agent* becomes a critical piece of the system, which depends on the reliability of a single agent and the communication lines that connect to it. As such agent should be updated with any relevant event that happens or even about intermediate results to be forwarded to other agents, the *Coordinator agent* might become a bottleneck in the information flow, affecting the performance of the whole system (that is, the response time of the system may be limited by the response time

---

[8]As Jennings explains in [104], one of the reasons that make agents in a MAS dependent on other agents is that no one individual agent has sufficient competence, resources or information to solve the entire problem, but they might have enough for some of the tasks to be taken.

[9]Axelrod's work on evolution of cooperation [9, 10] makes him one of the pioneers in such studies.

of the *Coordinator Agent*). In the worst case scenario when the *Coordinator Agent* collapses (as, for instance, it receives more requests and messages than it is able to manage in a given time span), the system may also completely collapse. One direct consequence of this fact is that, in real setups, a *Coordinator Agent* cannot be aware of all the relevant information but only a subset, reducing its capability to achieve optimal solutions. Another consequence is that centralized systems are hard to scale up, as the *Coordinator Agent* might collapse because of the growing costs in communication, synchronization and processing, all of them caused by an increasing number of messages coming from or going to more and more agents under the coordinator's command; and

- the other agents have a severe loss of *autonomy*, as the proper behaviour of the systems depends on the agents blindly accepting the commands of the coordinator.

An alternative is to distribute not only the work load but also the control among all the agents in the system (*distributed control*). That means to *internalize* control in each agent, which has now to be provided with reasoning and social abilities to make it able to reason about intentions and knowledge of other agents plus the global goal of the society in order to be able to successfully coordinate with others and also resolve conflicts once they arise. However, as Moses and Tennenholtz state in [144], in those domains where the cost of a conflict is dear, or if conflict resolution is difficult, completely independent behaviour becomes unreasonable. Therefore some kind of structure should be defined in order to ease coordination in a *distributed control* scenario. A good option taken from animal interactions is the definition of *social structures*.

1.2.1 Social Structures

Social structures[10], define a social level where the multi-agent system is seen as a *society* of entities in order to enhance the coordination of agent activities (such as message passing management and the allocation of tasks and resources) by defining structured patterns of behaviour. Social structures reduce the danger of combinatorial explosion in dealing with the problems of agent cognition, cooperation and control, as they impose restrictions to the agents' actions. These restrictions have a positive effect, as they:

- avoid many potential conflicts, or ease their resolution
- make easier for a given agent to foresee and model other agents' behaviour in a closed environment and fit its own behaviour accordingly.

Social structures are classified by Findler *et al.* in [70, 175, 109] into 5 groups: *alliances, teams, coalitions, conventions* and *markets*.

- An *alliance* is a temporary group formed voluntarily by agents whose goals are similar enough. The agents give up, while in the alliance, some of their own goals and fully cooperate with the other members of the alliance. Agents stay in the alliance

---
[10]They are also called *Artificial Social Systems* by Shoham, Moses and Tennenholtz in [144, 199]

as long as it is in their interest, thereafter they may join another alliance or stay on their own [109].

- A *team* is a group formed by a special agent (called the *team leader*) who recruits qualified members to solve a given problem.

- A *coalition* is similar to an alliance, as it is a temporary group where members do not abandon their individual goals but engage only in those joint activities whose goals are not in conflict with their own goals.

- A *convention* is a formal description of forbidden or preferred goals or actions in a group of agents.

- A *market* is a structure which defines two prominent roles (*buyer* and *seller*) and defines the mechanisms for transacting business.

Apart from these 5 groups of social structures, there is some work in the literature devoted not to define and formalize a special kind of social structure but to the emergence of such structures, mainly by using *referral networks* [108, 189, 174, 230] to model the social structures emerging from the interaction between agents, and using them to search for reliable sources of information or for services, as those networks implicitly build not only *referral chains* but also *trust chains*.

A more generic approach is proposed by V. Dignum *et al.* in [60]. In this case, instead of choosing a *sociological classification* (such as the Findler *et al.* one), the authors present a classification that comes from an *organizational theory* point of view. In this proposal social structures are divided in three groups:

- *Markets*, where agents are self-interested, driven completely by their own goals. Interaction in markets occurs through communication and negotiation.

- *Networks*, where coalitions of self-interested agents agree to collaborate in order to achieve a mutual goal. Coordination is achieved by mutual interest, possibly using trusted third parties.

- *Hierarchies*, where agents are (almost) fully cooperative, and coordination is achieved through command and control lines.

Table 1.1[11] summarizes some of the features that were taken into account in order to do the classification. The aforementioned organizational theory view endows this classification with two advantages:

- the three groups proposed by V. Dignum *et al.* aim to classify both human and software agent organizations. Therefore, this classification is more generic and, in the case of software agent organizations, it covers the five groups identified by Findler *et al* as well as referral networks. The relation between both classifications is shown in Table 1.2.

---

[11]This table summarizes tables 1 and 2 in [60].

	Market	**Network**	**Hierarchy**
Type of society	Open	Trust	Closed
Coordination	Price mechanism	Collaboration	Supervision
Relation form	Competition	Mutual Interest	Authority
Tone or climate	Precision/ suspicion	Open-ended/ mutual benefits	Formal/ bureaucratic
Conflict Resolution	Haggling (Resort to courts)	Reciprocity (Reputation)	Supervision

Table 1.1: V. Dignum *et al.* classification of Social Structures.

V. Dignum *et al.*	**Findler *et al.*** (extended)
Market	*market*
Network	*alliance, coalition, convention* and *referral networks*
Hierarchy	*team*

Table 1.2: Comparison among V. Dignum *et al.* and Findler *et al.* classifications of Social Structures.

- despite its generic nature, this classification is useful at the design stage, as it tries to motivate the choice of one of such structures based on their appropriateness for a specific environment. Thus, for each of the groups it identifies (*a*) the kind of problems that are well-suited to solve, and (*b*) some of the facilitation tasks the social structure should provide:

  - *Market structures* are well-suited for environments where the main purpose is the *exchange* of some goods. In this case the social framework proposed identifies three tasks to be performed by facilitator agents: *Matchmaking* facilities to keep track of the agents in the system, their needs and mediate in the matching of demand and supply of services; *Identification* and *Reputation* facilities to build confidence for customers and offer a certain degree of guarantees to all its members despite the openness of the system.

  - *Network structures* are well-suited for environments where (dynamic) *collaboration* among parties is needed. In this case one of the main facilitation tasks is the one of the *Gatekeeper*, which is responsible for accepting and introducing new agents into the society; *Notaries* are facilitator agents which keep track of collaboration contracts settled between agents, while *Monitoring agents* can check and enforce the rules of interaction that should guide the behaviour in the society.

  - *Hierarchical structures* are well-suited for environments where the society's purpose is the efficient *production* of some kind of results or goods or the control of an external production system. In these environments a reliable control of resources and information flow requires central entities that manage local resources and data but also needs quick access to global ones. In this scenario two main facilitation tasks are identified: *Controllers*, which monitor and orient the overall performance of the system or a part of it; *Interface agents* responsible for the communication between the system and the *outside world*.

1.2.2 Commitments and Conventions

Jennings in [104] presents an analysis of existing coordination models [176, 168, 221, 92, 122, 207] and proposes a unifying framework based on the concepts of *commitments* and *conventions*:

- *Commitments* are defined as pledges to undertake a specific course of action.

- *Conventions* are defined as the means of monitoring commitments in changing circumstances.

Jennings states that all coordination mechanisms can be reformulated by means of those concepts, in what he presents as the *Centrality of Commitments and Conventions Hypothesis*:

> *All coordination mechanisms can ultimately be reduced to (join) commitments and their associated (social) conventions.*

One important research issue is to find out how to introduce conventions in the interaction among agents. One way to do so is to code conventions as restrictions and hard-code them into the agents, to reduce the set of possible options to a limited subset. But that approach is against the idea of autonomy in agents. Another way is reducing the set of options at the protocol level. In this second approach the conventions to ensure coordination are subtly hard-wired into the protocols that the agents follow, in a way that the reaction of agents to a certain message is completely defined. Even though this scenario seems to be acceptable, as having the conventions hard-wired forbids the agents from deciding to violate the conventions, F. Dignum shows in [55] that such scenario has two drawbacks:

- *Autonomy loss*: agents always react in a predictable way to each message, as they follow the protocol. This predictable behaviour allows an agent to determine and even control the behaviour of other agents, as the agents would give response to a stimulus even if that goes against their own goals.

- *Adaptability loss*: if the protocols that they use to react to the environment are fixed, they have no ways to respond to changes in the environment (including violations of the conventions by other agents).

In [57], F. Dignum *et al.* propose that norms should be explicitly used as *influences* on an agent's behaviour. It should be noted that the proposal states that norms should not be imposed, they should only serve as a guidance of behaviour for agents, or agents will loose their autonomy. With this idea in mind, F. Dignum presents in [55] an agent architecture where social norms are incorporated to the BDI cycle[12] of the agent. In that architecture, norms are explicitly expressed in deontic logic and are divided in three levels:

- *Convention Level*: norms modelling the social conventions. At this level F. Dignum identifies two types of conventions: *interpretation rules* (which indicate how terms such as *reasonable* or *good* are defined in the context of this society) and *prima facie norms* (which describe general social norms and values). The latter are defined in terms of obligations of any agent a_i towards the whole society and are expressed in deontic logic as $O_i(x)$.

- *Contract Level*: norms regarding the obligations that arise when an agent a_i commits either to perform an action α or to achieve a situation p requested by agent a_j (the counterparty). Such commitment is modelled as an obligation of agent a_i towards agent a_j and is expressed in deontic logic as $O_{ij}(\alpha)$ or $O_{ij}(p)$.

- *Private Level*: to assure completeness of the model, the BDI model is extended to deal with obligations. At this level the intentions of the agent are seen as commitments towards itself to perform a certain action or plan, and are modelled as obligations of the form $O_{ii}(\alpha)$. By doing so, all the obligations the agent committed at any level are integrated inside the BDI cycle as a prioritized list of obligations that is continuously updated and which is used to decide the next actions to be performed.

This proposal is interesting, as it not only identifies different levels of commitment but also explains how such levels can be integrated in an extended BDI architecture. However the view of the proposed architecture is too agent-centered. Although some of the social conventions may be expressed at the *convention level*, no social structures are defined in order to ensure the proper behaviour of the society of agents as a whole[13]. So there is a need to find a way to model societies of agents whose interaction is regulated by means of some explicit norms.

1.3   Institutions

Most human interactions are governed by conventions or rules of some sort, having their origins in society (emergent) or the laws (codification of emergent rules) that societies have developed. Thus we find that all human societies, even the most primitive ones, have some kind of *social* constraints upon their members in order to structure and to regulate the relations among its members. Some of these constraints are quite informal (taboos, customs, traditions) while some others are formally defined (written laws, constitutions). Individuals integrate these rules into their behaviour so they become second nature, which is how the customs of one culture can create problems in the context of another culture. Modern human societies have defined collections of expected behavioural patterns that have an effect in lots of scenarios: shops, banks, conversations, lectures, clubs, etc.

---

[12]BDI stands for *Beliefs-Desires-Intentions* [176].

[13]An in-depth analysis of F.Dignum's approach is done in Section 3.2.2.

Institutions	Organizations
They are abstractions	They are (concrete) instances of institutions
Provide a structure for human interaction	Provide a structure for human interaction
Determine opportunities in a society	Are created to take advantage of those opportunities
Key concept: *norms*	Key concept: members' *strategies*

Table 1.3: Similarities and differences between Institutions and Organizations

1.3.1 Human Institutions

The economist Douglass North[14] has analysed the effect of this *corpora* of constraints, that he refers to as *institutions*, on the behavior of human organizations (including human societies). North states in [151] that institutional constraints ease human interaction (reducing the cost of this interaction by ensuring trust), shaping choices and making outcomes foreseeable. By the creation of these constraints, either the organizations and the interactions they require can grow in complexity while interaction costs can even be reduced. Having established these institutional constraints, every competent participant in the institution will be able to act—and expect others to act—according to a list of rights, duties, and protocols of interaction.

The main reason grounding the creation of institutions is to create trust among parties when they know very little about each other. No institutions are necessary in an environment where parties have complete information about others (e.g., a village market where vendors and buyers know each other and interact on a periodical basis). However, in environments with incomplete information (e.g., international commerce), cooperative solutions (based in trust) could break down unless institutions are created to provide sufficient information for all the individuals to create trust and to police deviations. [15]

1.3.2 Types of Institutions

Institutions can be created from scratch and remain *static* (e.g., the Spanish Constitution) or be continuously *evolving* (e.g., Common Law).

Institutions can be either *informal*, defined by informal constraints such as social conventions and codes of behaviour, or *formal*, defined by *formal rules*. Formal rules include i) *political and judicial rules* (constitution, statutes, common laws), ii) *economic laws* (common laws, bylaws) and iii) *contracts*.

---

[14]Douglas North received the Nobel Prize in 1993 thanks to his studies on the role of institutions in the performance of organizations.

[15]The same statements hold for closed multi-agent systems, where trust is implicit, and open multi-agent systems, where trust is something that has to be built by some kind of mechanism, such as the *electronic institutions* that are discussed in Section 1.3.3.

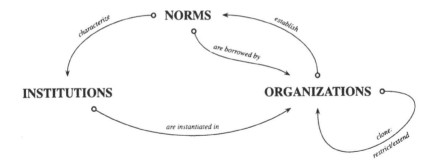

Figure 1.1: Some of the relationships between Norms, Institutions and Organizations

We will focus our attention on *formal institutions*. In such institutions the function of formal rules is to promote certain kinds of exchange while raising the cost of undesired kinds of exchange. Elnor Ostrom [159] classifies formal rules in 6 types:

- *position rules*: to define a set of positions (roles) and the number of participants allowed for each position,

- *boundary rules*: to state how participants are chosen to hold or leave a position,

- *scope rules*: to specify the set of outcomes and the external value (costs, induce-ments) related to them,

- *authority rules*: to specify the set of actions assigned to a position at a certain node,

- *aggregation rules*: to specify decision functions for each node to map action into intermediate or final outcomes,

- *information rules*: that authorize channels of communication among participants in positions and specify the language and form in which the language will take place (the protocol).

As norms[16] are, in fact, the elements that characterize institutions they are not only useful as norms to be followed, but also they enable people to recognize an organization (such as a particular company) as being an instance of a particular kind of institution, and then use this knowledge to predict other norms that could be applicable.

1.3.3 Electronic Institutions

An *Electronic Institution* (*e-institution*) is the model of a (human) institution through the specification of its norms in some suitable formalism(s). The statement is simple, but the practice is less so, partly because of the wide variety of norms and the different levels

---

[16]North does not differentiate between the terms *norms, constraints* or *rules*. From now on we will use the term *norm*, which is the more generic one, until we present a *full* definition of those terms in Chapter 4.

of abstraction in which they are defined, so that the form of expression (deontic logic, temporal logic, modal logic, classical predicate logic, types, for example) for one level is not necessarily suitable or even usable for another and partly because this stratification then leads to a need to demonstrate coverage and consistency between the levels (see [161] for a further discussion of these ideas). Nevertheless, the fact is that the essence of an institution, through its norms and protocols *can* be captured in a precise machine processable form and this key idea forms the core of the nascent topic of institutional modelling.

As we observed earlier, typically, many human institutions have evolved through millennia of interaction, and now we are facing the invention of new organizations (typically as business models) which are, in effect, human experiments with institutional structures. This practice is now being taken up, but with considerably more freedom of design, in the multi-agent systems community, where it is software agents that may come to the electronic institutions and which gain from a simplification of the interaction protocols and a confidence that a transaction will be properly concluded according to the norms of the institution in which they are participating. The aim is to build *e-institution* frameworks to formally design and implement agent architectures where interaction is ruled by some kind of institutional framework.

E-Institutions (as norm providers) are advantageous in the context of MAS because they:

- reduce uncertainty about other agents' behavior inside the institution,

- reduce misunderstanding, as there is a common corpus of norms governing agent interaction,

- permit agents to foresee the outcome of a certain interaction among participants,

- simplify the decision-making process inside each agent, by decreasing the number of possible actions.

However, one of the problems that the e-institutions field should address is notation. To give just an example, in 2001 there was an Agentlink meeting about Agent-Mediated Electronic Commerce, held in Prague[17]. One of the sessions was devoted to *e*-institutions. But different speakers used a variety of terminologies to refer to, sometimes, quite similar concepts:

- *norms, rules, policies* or *protocols*, to refer to the constraints for agent behaviour;

- *e-Institutions, Virtual organizations* or *Virtual Enterprises*, to refer to the implemented multi-agent architectures where agent behaviour is constrained;

- *human institutions* or *human organizations*, to refer to the systems in the human society that are modelled through *e*-institutions;

---

[17]A report summarizing all the sessions can be found in http://www.iia.csic.es/AMEC/ PragueSummary.html.

- *EI-Building* tools or *Meta-Institutions*, to refer to the tools and/or services to be provided in order to ease the design and/or implementation of agent-based electronic institutions.

Another terminology issue that could be observed was the use of the same concept (e.g., *norms*) with different meanings. Therefore some kind of unification is needed in order to ease the knowledge exchange among researchers.

We will review the different conceptions and terminologies used in the field in Section 3.1.3. Our proposed terminology and its meaning will be presented in Chapter 4 and Chapter 5. It is based in our view of the relationships among the terms *organization, institution* and *norms*, which is depicted in Figure 1.1. This expresses the idea that there is a collection of norms, which may be fairly abstract desires or wishes such as *fairness* or *equity*, such that particular sets of norms characterize institutions and then that organizations may either be instances of institutions or a combination of a bespoke set of norms. Such an organizational model may then be copied and expanded or simplified, while another organization may establish new norms which may be added into the general collection.

1.4   Summary

In this chapter we have presented an introduction to the *e*-institutions area and all the important aspects to be considered. First we introduced the problem of control *versus* autonomy, which specially appears in open multi-agent systems. Then we presented the concepts of *coordination, coordination theories, social structures, commitments* and *conventions*, to end with the definition of *human institutions* and their electronic counterparts, the *e-institutions*.

Our main observation is that *e-institutions* are advantageous in the context of MAS because they specify norms that:

- define what is forbidden (illegal) and what is permitted (legal) in an agent society;

- define the rights and responsibilities of the members;

- help agents to improve the achievement of their (socially accepted) goals, as agents can plan their actions taking into account the expected behaviour of the other agents.

In this book we propose a multi-level normative framework, deriving from and developing the notion of an electronic institution presented in Section 1.3.3, to define and police norms that guide, control and regulate the behaviour of the heterogeneous agents that participate in the institution. This framework aims to fill the gap between formal specifications of norms and the final implementation of a multi-agent system that should observe those norms. Doing so, we aim to bring together theoretical and practical approaches on norm and institutional modelling, extending the use of multi-agent systems in complex, highly regulated domains.

As application domain we will abandon the widely studied e-commerce field – which, most of times, can be modelled by sets of quite simple rules based in game theory– to focus on scenarios with complex normative frameworks such as international e-business or medicine. As a case study we will explore the possibility to formalize an electronic institution in a complex domain such as human organ and tissue transplantation [214]. Such real setup has to cope with:

- The (typical) problems that arise in medical applications. Those problems are caused by the ill-definition of the domain (is hard to foresee all the conditions that may occur in a patient and model them in a computer-readable representation). The ill-definition may result, for instance, in unexpected side effects (e.g., heart collapse) when certain decisions or actions (e.g., the prescription of a drug) are taken.

- The variety of regulations (national and international) that should be taken into account in the decision-making. Such regulations have to be modelled and included into the system.

Chapter 2

CARREL: an Agent-Mediated System for Organ and Tissue Allocation

In this chapter we present the formalization of *CARREL*[1], an agent-mediated electronic organization for the procurement of organs and tissues for transplantation purposes, as an electronic institution using the ISLANDER [68] institution specification language as formalizing languages. We show some aspects of the formalization of such an institution, using examples of fragments in the language used for the textual specification, and how such formalization can be used as blueprint in the implementation of the final agent architecture, through techniques such as skeleton generation from institution specifications described in [218] and [212]. At the end of the chapter the limitations the language imposes are shown and the need of more expressive and powerful formalisms is stated.

2.1 Introduction

The use of Multi-Agent Systems (MAS) in health-care domains is increasing. Such Agent-mediated Medical Systems are designed to manage complex tasks and have the potential to adapt gracefully to unexpected events. However, in that kind of systems the issues of privacy, security and trust are particularly sensitive in relation to matters such as agents having access to patient records, what is an acceptable behaviour for an agent in a particular role and the development of trust both among (heterogeneous) agents and among users and agents.

Organ and tissue transplantation are widely-used therapies against life-threatening diseases. But there are two issues that make transplantation management a very complex

---
[1]Upon Alexis Carrel [1873-1944], who received the 1912 Nobel Prize. He laid the groundwork for further studies of transplantation of blood vessels and organs.

issue: (i) *scarcity* of donors, so it is important to try to maximize the number of success-ful transplants (ii) *donor/recipient matching*, because of the diversity and multiplicity of genetic factors involved in the response to the transplant. In this chapter we propose an agent-based architecture for the tasks involved in managing the vast amount of data to be processed in carrying out

- recipient selection (e.g., from patient waiting lists and patient records),
- organ/tissue allocation (based on organ and tissue records),
- ensuring adherence to legislation,
- following approved protocols and
- preparing delivery plans (e.g., using train and airline schedules).

The relative scarcity of donors has led to the creation of coalitions of transplant or-ganizations. In the case of the United States of America, a new organization called UNOS (*United Network for Organ Sharing* [209]) has appeared in order to join and coordinate the several pre-existing transplant organizations that existed in some states. Also inter-national coalitions of transplant organizations have been created, such as EUROTRANS-PLANT [69] (Austria, Belgium, Germany, Luxembourg, the Netherlands and Slovenia) or Scandiatransplant [192] (Denmark, Finland, Iceland, Norway and Sweden). Indeed there is an initiative called *The Donor Action Foundation*[2] [63] which plans to create a world-wide coalition. This new, more geographically distributed, environment makes an even stronger case for the application of distributed software systems to solve:

- *the data exchange problem:* exchange of information is a major issue, as each of the actors collects different information and stores it in different formats. The obvious, and easily stated, solution is the definition of standard data interchange.

- *the communication problem:* countries typically use different languages and termi-nologies to tag the same items or facts. Either a standard notation or a translation mechanism needs to be created to avoid misunderstandings.

- *the coordination issues:* in order to manage requests at an international level, there is the need to coordinate geographically distributed surgery teams, and to coordinate piece[3] delivery at an international level.

- *the variety of regulations:* an additional issue is the necessity to accommodate a complex set of, in some cases conflicting, national and international regulations, legislation and protocols governing the exchange of organs. These regulations also change over time, making it essential that the software is adaptable.

The consideration of the factors that above we have summarized leads to the ques-tion of whether some kind of automation of the assignation process is desirable and if so,

---

[2]The Donor Action Foundation is an initiative of the Eurotransplant International Foundation, the Spanish National Transplant organization (ONT) and the former Partnership for Organ Donation (U.S.A.).

[3]We will use the term *piece* to refer to *tissues* and *organs*.

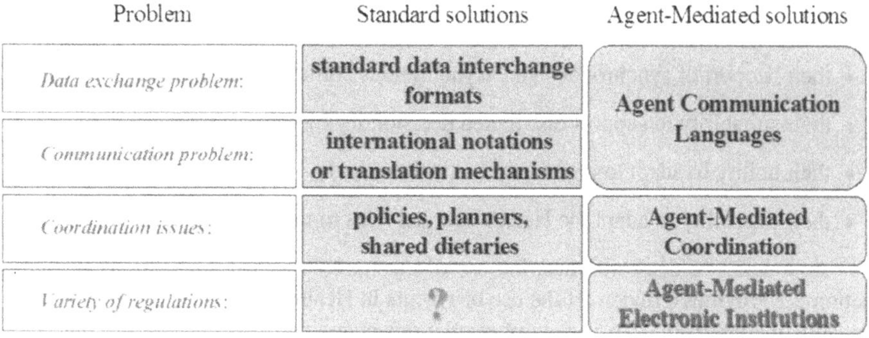

Problem	Standard solutions	Agent-Mediated solutions
Data exchange problem:	**standard data interchange formats**	**Agent Communication Languages**
Communication problem:	**international notations or translation mechanisms**	
Coordination issues:	**policies, planners, shared dietaries**	**Agent-Mediated Coordination**
Variety of regulations:	**?**	**Agent-Mediated Electronic Institutions**

Figure 2.1: Problems in a distributed organ and tissue allocation system

whether it is possible. The first two points can largely be resolved using standard software solutions. In fact most of the work in the field of transplant allocation (such as EU projects RETRANSPLANT [186][4], TECN [204]) is devoted to the creation of a) standard formats to store and exchange information about pieces, donors and recipients among organizations, b) telematic networks, or c) distributed databases. Project ESCULAPE [67] uses conventional software to help in tissue histocompatibility by developing HLA[5] referencing computer systems and software packages to be used by hospitals and laboratories as a human tissue typing tool. However none of them give support to the coordination and regulatory problems, which should be solved entirely by humans.

The third point (coordination) is harder to solve with conventional software, as coordination tasks need to have the appropriate domain knowledge to be properly carried out. However, medicine is one of the most difficult fields for coordination, as it is extremely difficult to foresee all the conditions that may occur, leading to unexpected side effects when certain decisions or actions are performed in a unanticipated situation. As Fox & Das argue in [76], these are the kind of ill-defined fields which have historically been the concern of Artificial Intelligence. They also identify software agents as having many strengths (mainly *pro-activity* and *autonomy*) which make Agent Technologies well-suited for medical applications. This view is also supported by Shankararaman *et al.* in [197], with also gives some arguments of the use of agents in medical applications.

---

[4]Unfortunately, there is no information available about the practical results of those projects other than the project URL.

[5]HLA stands for, *Human Leukocyte Antigen* system, a group of the most important antigens responsible for tissue compatibility, together with the four significant genetic markers (on chromosome 6) that encode them (HLA-A, HLA-B, HLA-C and HLA-D). The HLA system is used as one of the tests in matching donor and receptor tissues or organs in the allocation process. The test is composed by six compatibility factors, and current policies consider that a match of three or more of these factors is needed in order to consider compatibility between donor and recipient.

Such arguments can be summarized in:

- the capability of agents to anticipate pro-actively the information needs of users,

- their support of synchronous and asynchronous communication among parties,

- their suitability to support distributed decision making,

- their ability to adapt to unpredicted situations, and

- their capability to adapt the Health Care services to the patient needs

As a result of these abilities, the use of agents for medical applications is rising. Section 2.2 provides a review of the use of Agents in Health Care.

It is the last point (the variety of regulations changing over time) which underpins our case for the use of so-called *e-institutions*, whose purpose is to provide over-arching frameworks for agent interaction, where agents may reason about the norms, in the same way as physical institutions do in the real world through social norms. E-Institutions and the norms that govern them are the key to a system that is able to adapt automatically to changes in regulations. These norms define the *acceptable* actions that each agent may perform depending on the role or roles it is playing, and clearly specifies the data it may access and/or modify in playing those roles.

2.2 Medical Informatics and Agents

For a long time, experts in Computer Science have been interested in the medical domain. This interest comes no only from the social value that technological applications may create (by improving health care services or medical knowledge), but also from the implicit complexity of the domain. Such complexity makes the medical domain an interesting test case for probing and tuning modeling techniques, algorithms and architectures.

Medical Informatics (also known as *Electronic Health Care* or *e*-Care) is the field of Computer Science devoted to find applications in Information Technologies in order to:

- Create, store, restore and distribute information about patients

- Define, select and improve the diagnosis and treatment procediments

- Represent, apply and obtain medical knowledge

- Establish communication and collaboration among patients and medical practitioners

- Support the management and organization of Health Care services.

One of the research fields that has had more interest in medicine-related problems is Artificial Intelligence. Medical Domain has been always a fertile area for Artificial Intelligence applications. In the last 30 years sophisticated knowledge-based systems have been introduced, such as MYCIN [200] or, more recently, CAPSULE [220]. The focus in such systems is given to Knowledge Representation and, for distributed systems, to Ontology construction. There are several Internet resources for medical terminology[6], but one of the most recent and interesting efforts in Knowledge Representation for distributed systems is GRAIL [184]. There also exists a lot of literature on applying different techniques, tools and available knowledge to enrich the reasoning process of those systems (see for example [86]), or to apply other techniques (such as Natural Language and Robotics) to develop assistive technologies that interface with the patient (see [141]).

The raise of what is called the *Information Society* joined with the need of promoting Health services (mainly in the United States of America, where Health is just another market) are the origin of a change in the way the health service providers are managed and presented, from a institutional-centered approach to a patient-centered one, in order to create a personalized environment to attract patients.

This new situation has led some research groups to think in Software Agents as a feasible technology which provides solutions to some of the problems and objectives the Health Care sector is interested in. As explained in [197], Agent Technology allows:

- To proactively anticipate the information needs of a patient, and deliver it in a periodical basis

- To support communication and coordination, either synchronous or asynchronous, among all members of a medical team, allowing the share of distributed information and knowledge sources, and providing distributed decision making support.

- To adapt medical services to patients' needs (*personalization*).

- Increase the patient's control over all the data collected in his/her medical records.

With all that in consideration, some agent-based systems have been developed in order to solve some problems. According to Shankararaman *et al* [197], the agent applications in the medical area can be classified in four groups:

- Management of patient-oriented information.

- Cooperative management of patients.

- Autonomous patient monitoring and diagnosis.

- Remote Health Care services.

2.2.1 Patient-Oriented Information Management

This kind of applications aims to personalize the information delivered and the services offered to patients. To do so, each patient has *associated* one or several agents which have access and manage the patient's medical records. With such information they can not only filter and adapt the information delivered to the patients, but also they can be *proactive*, foreseeing the patient needs and autonomously seeking for information or services that might be helpful for the patient.

---

[6]A good compilation of Internet resources for medical terminology can be found in [102].

An example of this kind of systems is R2Do2 [201]. In this system users receive information as recommendations ("ToDos"[7]) which are generated by a set of agents from a knowledge in the medical domain that medical practitioners have introduced previously in the system.

2.2.2 Cooperative Management of Patients

This application area studies ways to optimize the management of health services offered by the centers to the population. The application of multi-agent systems is focused in the development of distributed management and coordination of medical teams, facilities and all the related resources.

There are some examples of this kind of applications. GAMES II [111] is a multi-agent system where diagnosis is distributed among several agents, each one specialized in one kind of diagnosis (such as those related with blood or the heart). In order to allow the agents to use the information created by the other agents, an ontology has been created to define a common semantics to ensure a mutual understanding among the agents in the system. Another interesting example is ADDCare [100], a prototype of an agent architecture for a) decision-making with a certain grade of uncertainty, and b) management of tasks to be done. The proposed architecture is structured in three levels: the *domain level*, the *inference level* (where reasoning is performed) and the *control level* (which has meta-rules working on the inference level).

Some systems are focused on applications of planning methods. The work of Decker [52] on the design of coordination mechanisms for groups of agents applied to Hospital scheduling is a good example. Miksch's work [140] is also a good example of planning in the medical domain.

There are other applications that are focused in the implementation of medical protocols (also known as *guidelines*), such as the work of Pattison-Gordon *et al* [166] or the work of Fox *et al* [77]. A related approach for monitoring medical protocols is described by Alsinet *et al.* in [4].

2.2.3 Autonomous Monitoring and Diagnosis of Patients

This kind of applications provide the agents with all the domain knowledge needed, not only to allow the agents to identify important changes in the state of a patient (in the monitoring cycle) but also to make the agents capable to perform all the reasoning to autonomously provide a diagnosis.

Another interesting example is GUARDIAN [112], a multi-agent system supports monitoring and diagnosis for patients during cardiac post-operatory periods. The system monitors the signals of the sensors attached to the patient, autonomously identifying dangerous states and providing the caregivers not only with an alarm but also with a diagnosis of the problem.

---

[7]The term comes from the verb "to do".

These kind of systems that aim to build diagnosis without the intervention of human practitioners have to deal with legal issues, though. There is a lot of discussion, not only in the medical domain but also in other critical areas about the feasibility of letting agents to do diagnosis and act according to such diagnosis without supervision of a human expert. Those discussions not only are related to the soundness of the reasoning involved in such diagnosis but also in legal figures: who should be blamed when a patient is hurt due to a wrong diagnosis? [75]

2.2.4 Remote Health Care Services

This kind of systems aim to use Internet and other communication networks (such as optic fiber) that are currently arriving to the patients' homes in order to provide remote health care services.

These services are specially useful for people with reduced mobility, such as handicapped and/or senior citizens, allowing them to receive the information and the services they need in their own homes. A good example is [29], which proposes a multi-agent architecture combining mobile and non-mobile agents for 1) monitoring the patient's state, 2) remind him/her when a drug has to be taken, 3) assist the patient in some of the daily tasks, such as *buying food*, which is done electronically through Internet or other means. An extension of such proposal is presented in [28], where the idea of deploying a virtual community for elderly support is presented.

An example of an application that is currently used is INCA [13], a system that integrates the health care services available in a city. Coordination and planification is achieved through agents negotiation in order to, for instance, coordinate the visit of practitioners to the patients' homes. The monitoring system uses domotic technologies to install monitoring devices in the patient home that are able, without the patient's intervention, to trigger an alarm and send it to the health care providers.

2.2.5 Agents and the Transplant Domain

There are very few references in the literature about the use of agents in the transplant domain. In [210] Valls *et al.* describe an agent that uses multi-criteria decision techniques in the selection of the best receiver in a transplant, providing the Hospital Transplant Coordinator with a result according to the weights the user assigned to each criteria. Moreno *et al.* present in [143] a hierarchical multi-agent system where the agent on the root node plans transport routes between hospitals using the information obtained from the other agents in the hierarchy, removing routes that will exceed the maximum available time for transportation and avoiding potential fatal delays due to mistakes in coordination of different means of transport. In [142] Moreno *et al.* propose a multi-agent system architecture to coordinate hospital teams for organ transplants. Coordination is achieved through agents that keep track of the personnel schedules and the availability of the facilities (both described as time-tables divided into slots of thirty minutes). Finally, Aldea *et al.* present in [3] an alternative design for a multi-agent architecture for the Spanish organ allocation process. It identifies the agents needed to solve the problem and organizes them in four levels (*Hospital Level*, *Regional Level*, *Zonal Level* and *National Level*). However, no formalism is used in the development of the architecture.

As far as we know, the *CARREL* system that we describe in the following sections is the first agent-mediated electronic institution applied to the transplant allocation problem that joins the strengths of *agents* with the advantages of formal specifications.

2.3   The Organ and Tissue Allocation Problem

Organ transplantation from human donors is the only option available when there is a major damage or malfunction in an organ. It is also very important in economic terms. For instance let us take the case of Spain, where transplantation of one kidney compared with dialysis would save between 186400 and 240530 Euros.

Over the years, transplant techniques have evolved, knowledge of donor-recipient compatibility has improved and so have immunosuppressant drug regimes, leading to an increase in the number of organs that can be transplanted, but also in the range of transplants, moving beyond organs (heart, liver, lungs, kidney, pancreas) to tissues (bones, skin, corneas, tendons). However, the allocation process for tissues is quite different from that for organs, because of the time such pieces can be preserved outside the human body. Tissues are clusters of quite homogeneous cells, so the optimal temperature for preservation of all the cells composing the tissue is almost the same. Thus, tissues can be preserved for several days (from six days in the case of corneas to years in the case of bones) in tissue banks. For tissues, the allocation process is triggered when there is a recipient with a need for a certain tissue, at which time some number of tissue banks are searched for a suitable one.

Organs, on the other hand, are very complex structures with several kinds of cell types with different optimal preservation temperatures. That fact leads to quite short preservation times (hours), no need for an organ bank, and an allocation process that is triggered when a donor appears, taking the form of a search for a suitable recipient in some number of hospitals.

Since 1980 the number of requests for the application of transplant techniques has risen so much[8] that the human coordinators—the people at the hospitals who act as the interfaces between the surgeons internally and the organ and tissue banks externally—are facing significant problems in dealing with the volume of work involved in the management of requests and piece assignment and distribution. Transplant-based therapies are the subject of much investigation and increasing application, such that demand for pieces may well rise rapidly in the near future.

2.3.1   Coordination Structure: the Spanish Model

At the time of writing, more than one million people in the world have successfully received an organ, and thereafter, in most cases, lead normal lives. Spain has become the

---

[8]The continuous raise in requests is due, among other factors, to the introduction of new immunosuppressors which have significantly decreased rejection in recipients' clinical evolution.

first transplant organization in the World with 33 cadaveric organ donors per million population (pmp) in 1999 (37 donors pmp in Catalonia). This success places the Spanish Organización Nacional de Transplantes[9] (ONT) and the Catalan Organitzaciò CATalana de Trasplantaments[10] (OCATT) transplant organizations among the most effective and demonstrate the highest global volume of transplants per head of population, thanks to the creation and implementation of a network of well-coordinated hospitals and tissue banks, coupled to the definition of clear procedures for the distribution of organs and tissues and a high level of citizen awareness of the value of donation.

The ONT is a technical organization under the authority of the Ministry of Health and Consumer Affairs, without powers of direct management and without specific executive power. The ONT acts as a service agency for the whole National Health System, works for the continuing increase in the availability of organs and tissues for transplantation and tries to ensure the most appropriate and correct distribution of pieces.

The success of the Spanish model comes from a re-structuring and optimization of the process at two levels:

- *intra-hospital level*: where the role of Hospital Transplant Coordinator was created to improve the coordination of all the people working at any step of the donor procurement, allocation and transplantation process

- *inter-hospital level*: where an intermediary organization—OCATT for Catalonia, ONT for the whole of Spain—was created to improve the communication and coordination of all the participating health-care organizations, namely hospitals and tissue banks.

The ONT is now structured as a network system established at three basic levels: national, regional and local.

National Coordination

The National Transplant Coordinator is the head of the ONT, and has the mission to act as a nexus among a) local, national and European health authorities, b) health professionals, c) the different social agents involved in organ donation and transplantation and d) the general population. The National Transplant Coordinator works at the Central Coordination Office, which functions are listed in table 2.1.

Regional Coordination

Each of the seventeen Spanish Autonomous Communities has a Regional Transplant Coordinator, ehic are responsible for the coordination of resources, tasks relating to information, circulation and promotion at regional level. These regional coordinators also are members of the Organ and Tissue and Transplant Commission, where any subject related with transplantation that affects more than one Autonomous Community is discussed.

---

[9] National Transplant Organization
[10] Catalan Transplant Organization

Coordination	• Inter-hospital Coordination of all multiorgan retrieval procedures. • Up-date and maintenance of liver, heart and lung transplant waiting lists. • Cooperation in kidney exchanges. • Coordination of air/land transportation of transplant teams and organs for transplantation. • Cooperation in patients transfer if needed. Channeling of patients reports for pre-transplant evaluation. • Channeling of requests for bone pieces or other tissues. Channeling of "Bone Marrow Donor Searches".
Regulations and Reports	• Elaboration of any technical report related directly or indirectly with the organs, tissues and haematopoietic progenitors transplantation. • Promotion of Agreements and Consensus Reports
Studies	• Collection of data on procurement and transplantation activities. Data analysis. Publications. • Evaluation of health requirements: legal, human and material. • Promotion and coordination of multi-center studies and research projects
Information	• About donation and transplantation activities and health related topics to i) Health Administration, ii) transplant coordinators, iii) transplant professionals, iv) international transplant organizations. v) patients associations. • Information to the general public by means of i) public campaigns for social sensitization, ii) issuing of donor cards. iii) management of a telephone line to provide information about any question related with donation and transplantation • Spreading of informative, didactic and working material.
Others	• International Cooperation. • Promotion of Specific Training Courses. • Development of the Spanish Society of Donation and Transplantation of Organs and Tissues.

Table 2.1: ONT: functions of the Central Office (source: ONT)

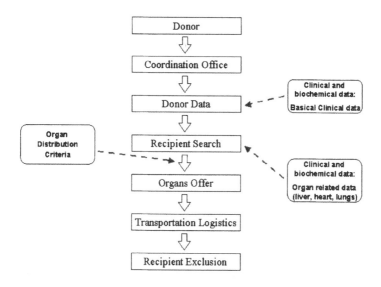

Figure 2.2: Donation alarms management.

Hospital Coordination

It is the hospital coordinators who catalyse the detection of donors. A full review of the hospital coordinator's role and the difficulties faced is presented by López-Navidad in [126]. In most cases this work is combined with their daily health care work so that the coordinators still maintain contact with the actual hospital life.

2.3.2 The Organ and Tissue Allocation Process

Emergencies

In Spain, the ONT manages a list of urgent cases (called *urgency 0*) containing all the recipients whose condition is life-threatening. If there is a suitable recipient for an organ in this list then he/she has higher priority in the assignation process over all other recipients.

Procurement

In the case of organs the process starts at the *procurement stage*, when the members of the coordination team inside a certain hospital are made aware of a potential donor. A donor alarm is then sent to the ONT—except in Catalonia, where it is sent to OCATT. This alarm is signalled by telephone, and a human member of the staff has to list the basic attributes of the donor, including the results of clinical analysis, and a first evaluation of the organs and tissues that could be extracted is carried out. This first call is done as early as possible, usually when brain death of the potential donor is diagnosed. Enough time is then available to organize the supply infrastructure and transport. Figure 2.2 depicts the process.

After carefully recording the donor's data, a dossier is opened for each case including an incident log sheet used to record all the steps taken and the time when each of then happens.

Search

The next step is to search for suitable recipients. The intermediary organization (ONT, OCATT) carries out a recipient search for each organ that may be available by calling all hospitals with information about the pieces. To speed up this search process, each organ is assessed separately with reference to the distribution criteria. If there is any emergency "0", it becomes in a national priority. In the rest of the emergencies the pre-defined distribution criteria are strictly applied. These criteria are divided into clinical and geographical criteria. The clinical criteria are established and reviewed every year by all the transplant teams and ONT representatives, whereas the geographical distribution criteria are established by the Interterritorial Council of the National Health System. In that way, Spain is divided into six areas (figure 2.3)).

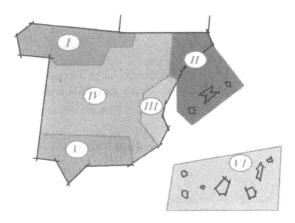

Figure 2.3: Zones defined in Spain.

Each day the turns for each area are established and the offers are made following a distance criteria: first the generating centre, otherwise the centres nearby in the same city or Autonomous Community, or otherwise the area (see figure 2.4). If an adequate recipient has not yet been found, the coordinator of the zone starts a general turn which works like a system of clogged wheels (figure 2.5). Each time that a team transplants an organ due to its turns area it goes to the last place of the area's wheel and the same occurs when transplanting an organ from the general turn. So, every day the turns of the areas

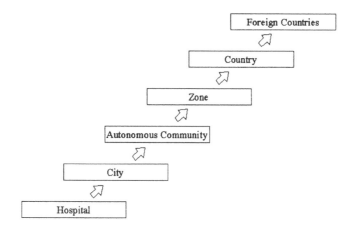

Figure 2.4: Distribution criteria for organs since 1996.

and the general area are changed depending on the transplants, that took place the day before.

In the situation that there is no suitable recipient in the country, the organ will be offered, through the OCATT, to other states and European transplant organizations.

Acceptance and Delivery

Once the most suitable recipients are located on the transplant centers' list, the offer is made to the transplant team through the hospital coordinator. All the donor's data are provided together with the conditions established by the generating hospital, particularly with regards to the time of removal and other possible requirements. The team, that is to perform the implant makes the final assessment, analysing all the information available about the organ, and decides whether the removal and implant can in fact be performed. If the offer is turned down, it goes to the next hospital in the area or otherwise to the general turn.

If the offer is accepted then the *delivery stage* starts: the generating hospital is informed, transport is organized (ambulance or helicopter from one hospital to one nearby, to the airport or to a train station; plane and/or train from one city to the other) and the main process schedule is defined, including the delivery plan (which must take into account transportation system schedules). The organization of transport depends on the distance to be covered and if it a removal team has to be mobilised or just the organ has to be carried.

- *Local donor.* A donor who is in the same city as the extracting/implanting team, but in another hospital. In this case the hospital coordinator organizes the transport according to the internal agreements among both hospitals.

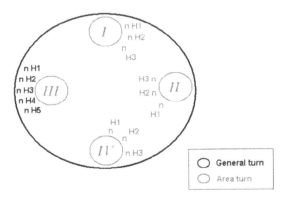

Figure 2.5: Turns inside each zone.

- *Non-local donor.* If the donor and the extracting team are in different cities, ONT staff will organize the transfer and that always involves a more complex operation. There are two scenarios: a) For distances less than 200 km., teams are usually carried in ambulances or helicopters. If necessary the collaboration of the police is requested to open the way or of the army is asked to help with military air transport and the use of landing bases. On occasions military helicopters are used and on other occasions they are civilian, normally belonging to the civil protection services provided by the Autonomous Communities. These means of transport are used as long as the weather conditions and schedule permit. b) For long distances, given the short period of physical ischemia that is tolerated by the organs, private aircraft are used for this type of distances and occasionally the help of the Air Force is required. At this point, it should be taken into account that the preparation of a flight requires at least two hours, (checking the aircraft, calling the crew, establishing the flight plan, etc), which is why it is so important to advise the ONT of the existence of a donor as soon as possible. When the flight plan is ready both hospitals are informed of the schedule, the company and the flight number.

Final Report

Once the removal team arrives at the hospital, the ONT staff waits to be informed of the implant so that the patient is immediately removed form the waiting list.

The hospital coordinator from the generating hospital will later send the ONT the donor's registration sheet duly completed with all the transplant performed.

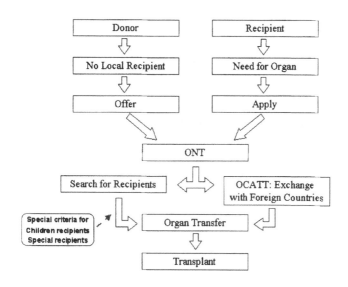

Figure 2.6: Management of kidney donations.

Renal Distribution

In Spain there are 40 centres that carry out kidney transplants, 6 of them pediatric hospitals. Each centre has a waiting list including its own patients and patients who receive treatment in other dialysis units for which the transplant team is a reference centre.

Each transplant hospital has its own transplant programme, the organs that are generated within the hospital and in hospitals that sends them transplant patients. Kidneys are distributed according to strict local criteria.

The central ONT office collaborates when necessary in exchange agreements, providing then with the infrastructure required. The ONT participates in the search for recipients for kidneys with special characteristics, usually when a recipient can not be found on the local waiting list (figure 2.6) or when a HLA typing unit detects that a kidney may be suitable for a hyperimmunized patient. For this purpose the ONT constantly updates the list of hyperimmunized patients and also distributes serum samples periodically exchanged between the immunization centres.

Tissue Distribution

In the case of organs, time is one of the key issues, since they can only be kept outside a human body for a short time-span (hours), as all the preservation methods only can delay, but not stop, the decay process arising from their not receiving enough blood (ischaemia).

In the case of tissues, such as corneas, skin or bones, they can be preserved for longer periods—days, even weeks. Such relative resilience permits not only an effective preservation process, but also allows time for assessing the quality of the tissues and establishing the absence of bacteria and viruses. The consequence is a search stage that works in the opposite way to that for organs. With organs, the process is triggered when a donor appears, whereas in the case of tissues, it is the appearance of a new recipient that triggers the search through the tissue banks for a suitable piece for transplant. This search too is done by means of several time-consuming telephone calls from hospitals to tissue banks.

2.4   The *CARREL* System

The aim of the *CARREL* system is to assist specialists in the decision-making during the allocation and distribution of pieces for transplants, in a manner that is acceptable according to the legislative requirements and other procedures governing the process. To achieve this, the agents composing the *CARREL* system have to be given with the appropriate domain-specific knowledge (kinds of pieces, attributes to describe them, etc.) so they can act *rationally*[11], and also with the rules they should follow, such as which actions can be done when, what information can be accessed or given out, etc.

2.4.1   Analysis of the System

We chose to use Agent Mediated Electronic Institution and regular multi-agent systems to automate part of the allocation process, finding suitable pieces for transplant and giving support to the decision making steps in that process. Some of the agents in the system should also be able to create, negotiate and coordinate plans for the extraction, transfer and implantation of pieces. The over-arching goals are not only to assist in the process of selection and procurement but also to make an important improvement in the results of the process, that is, to optimize the allocation based on time and compatibility constraints.

As guidance for our work, we followed the Spanish model of organ allocation described in Section 2.3, as it is considered one of the world's most successful organization. Our proposed system follows the same two-level optimization that was developed in Spain when the transplant management system was created. So there are agent solutions for the *inter-hospital* and *intra-hospital* levels. However, in order to exploit the fully potential of a distributed agent architecture, some parts of the process have been optimized in order to reduce the *seek time*, that is, the time needed to find a proper *recipient* for a given *piece*.

The Inter-Hospital Level

At the inter-hospital level we have created the *CARREL Institution*, an agent platform which hosts a group of agents (an *agency*) responsible for the allocation of organs and tis-

---

[11] In the sense of Simon and Newells' Bounded Rationality [148].

sues. In this agency different entities (the agents) play different roles that are determined by their goals, rights and duties.

In the case of tissues, the allocation process comprises:

1. The tissue banks keeping the institution updated about tissue availability

2. The agency receiving requests from the hospitals for tissues. For each request (brought by an agent representing the hospital) the institution tries to allocate the *best* tissue available from all the tissue banks that are known.

In the case of organs, the process comprises:

1. Each hospital informing the institution about patients that have been added to or removed from the waiting list of that hospital, or patients either to be added to or removed from the national-wide Maximum Urgency Level[12] Waiting List.

2. When a donor appears, the hospital informs the institution of all the organs suitable for donation in the form of *offers* sent to the organ allocation organization, which then assigns the organs.

Figure 2.7 depicts all the entities that interact with the *CARREL* system, where TB denotes a tissue bank and UCTx denotes a transplant coordination unit. There are a) the hospitals that create the piece requests, b) the Tissue Banks, and c) the national organ transplantation organizations, that own the agent platform and act as observers—the figure shows the organizations in Spain: the Organización Nacional de Transplantes[13] (ONT) [155] and the Organitzaciò CATalana de Trasplantaments[14] (OCATT). We make it a requirement in our model that all hospitals, even the ones that own a tissue bank, will make their requests through *CARREL* in order to ensure an acceptable distribution of pieces and to ease the tracking of all pieces from extraction to implant, in the same manner as ONT and OCATT require for organs.

The role of the *CARREL* Institution can be summarized in terms of following tasks:

T1 to make sure that all the agents which enter into institution behave properly (that is, that they follow the behavioral norms).

T2 to be up to date about all the available pieces in the Tissue Banks, and all the recipients that are registered in the waiting lists.

T3 to check that all hospitals and tissue banks fulfill all the requirements needed to interact with *CARREL*.

T4 to take care of the fulfillment of the commitments undertaken inside the *CARREL* system.

T5 to coordinate the piece delivery from one facility to another.

T6 to register all incidents relating to a particular piece.

---

[12]In Spain the Maximum Urgency Level is called Urgency-0
[13]National Transplant organization
[14]Catalan Transplant organization

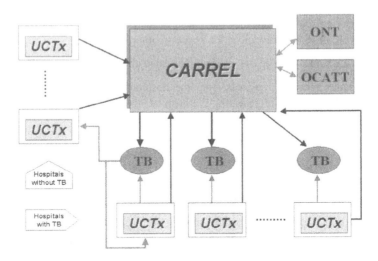

Figure 2.7: The *CARREL* System environment: the *CARREL* Institution and the UCTx

A hospital becomes a member of the *CARREL* institution in order to make use of the services provided. In doing so, they accept to respect the norms that rule the interaction inside *CARREL*. Some of these norms are:

N1 All organ offers and tissue requests should be done through the *CARREL* institution.

N2 Hospitals must accept the outcomes of the negotiation (assignation) process.

N3 Hospitals receiving an organ or tissue from *CARREL* must update the institution with any relevant event related to these organs and tissues.

A hospital is represented in *CARREL* by the Transplant Coordination Unit (*UCTx: Unidad de Coordinación de Transplantes*). This agency serves as interface between the surgeons and *CARREL*. When a surgeon needs a piece she makes her request through the UCTx system, which analyses the information entered by the surgeon, adds the information about the recipient and, finally, creates a *Finder Agent*, that is, the agent that goes to *CARREL* looking for a suitable piece. Appendix B has a complete description of the UCTx multi-agent system.

The information required by the *Finder Agent* to look for a piece in *CARREL* is held in an electronic *Sealed Envelope*. The information contained in the envelope is summarized in Table 2.2. The *Selection Function* is the part of the information contained in the *Sealed Envelope* that allows the *Finder Agent* to perform a negotiation. It is composed of a set of rules, each one a constraint on the piece to be selected. Some of these rules may originate from the policy of the whole transplant unit of the hospital, but the others are introduced by the surgeon, who can set the constraints associated with a given recipient. Table 2.3 lists the kind of predicates a *Selection Function* can include.

Urgency level	that works as electronic postage stamp and sets the urgency level of the request (in Spain: normal, urgency-1 or urgency-0)
Hospital identification	a certificate issued by the Certification Authority associated with the *CARREL* institution, to allow the institution to authenticate the sender of each request and ensure that only *Finder Agents* with requests from authorized hospitals can enter and negotiate inside *CARREL*.
Piece information	type, parameters, etc.
recipient data	age, sex, laboratory analysis, etc.
selection function	see Table 2.3

Table 2.2: The envelope contents

piece	predicates that describe the constraints the selected piece has to satisfy, such as the age of the donor or the dimensions of the piece itself.
origin	predicates that can set constraints about the tissue bank(s) preferred by the surgeon or the hospital
cost	predicates about the cost of the piece, such as price. Note: the cost is just that of extraction and preservation. Settlement is managed via a clearing house.

Table 2.3: The Selection Function predicates

The Intra-Hospital Level

The functioning of the Transplant Coordination Unit (UCT) may also benefit from the use of agents to help coordinate all the people in the hospital related to a particular case. Hence, our modeling of the UCT represents not only the surgeons, but also:

- the human transplant coordinator, who is the overall coordinator and who must coordinate all the tasks to be done and who must also be informed of any transplant related event

- any member of the hospital staff who plays a role at any step of the transplant process, from the moment when the piece arrives to the hospital until the piece has been implanted in the recipient.

2.4.2 Important Requirements

Rule-Guided Behaviour

In this kind of medical application the use of rules in order to guide the agents' behaviour is mandatory, as any mistake can lead to an unsuccessful transplant and potentially the death of a patient, as well as the waste of a piece which might have better benefited someone else. A further complication is that, as above mentioned, all the agent actions must respect legislation on the distribution and use of pieces for transplantation. As we explained in Section 1.3.3, electronic institutions work with explicit representations of norms [56, 36, 46, 53]. Expressing all the regulations and protocols in the form of computable norms—instead of hard-coding them so they are scattered throughout the logic of a program—not only admits a readily verification, both informal and formal, of adherence but also it gives the system the added flexibility of behaviour that it can be adapted in the light of regulatory changes (an event that is not uncommon).

Security Considerations

Transplant information is considered high-risk data as it includes sensitive information about people (donors and recipients). So the UCTx and *CARREL* systems have to observe the local, national and European Union legislation on transplants (see the reports of the ONT in [158] and the recommendations of the Transplant Experts Committee in [132]). It also should follow the European directives and the Spanish Law on personal data protection [118] [62] [182]. In particular, both have to ensure confidentiality, privacy and integrity of patient and donor data. This is a long-standing issue in health care that acquires new facets with the use of Electronic Medical Records (EMR). One of the benefits of using electronic records is that it assures access for authorized and authenticated users as well as tracking access as demanded by law. It is important for patient and donor trust that it can be demonstrated that the information about them may not be used for any purpose beyond that for which it was collected.

Although security is an important issue, we will not discuss the details of the security measures on this chapter. In [27] we already made a report with the security safeguards to be included in a multi-agent system such as *CARREL*. In Appendix A a review can be found on the security measures that Spanish and European Law impose to any system managing medical records.

Formal Approach

In [76, 75] Fox and Das state that there is a need to develop methodologies and, if possible, formal methods (formal specification, refinement and verification) when applying agents to the medical domain in order to improve safety by improving the integrity of the agents' internal code and avoiding *ad hoc* practices. Fox and Das also present the PRO*forma* formalism, which permits the definition of guidelines to be followed by each agent in the system. For our purposes and from the point of view of distributed systems, the problem is that this formalism does not cover all communication acts and the dependencies between guidelines and hence how the behaviour of one agent may affect another. It also includes no definition of *roles* in its formalism.

Because of the need to model our system in a way that was accurate and verifiable, we searched for an agent-based methodology that included an *e*-institutions formalism which could be applied to real setups. At the time we started working in *CARREL*, The ISLANDER formalism [68] appeared (and it still appears) as a sound option, as it not only had a serious work on formalization [149, 188] but it has also shown it is well-suited to model practical setups (electronic auction houses). It also attracted us the fact that a full suite of design tools was in development.[15]

The ISLANDER formalism views an agent-based *e*-institution as a type of *dialogical system* where all the interactions inside the institution are a composition of multiple dialogic activities (message exchanges). These interactions (that Noriega calls *illocutions* [149]) are structured through agent group meetings called *scenes* that follow well-defined protocols. This division of all the possible interaction among agents in scenes allows a modular design of the system, following the idea of other software modular design methodologies such as the Modular Programming or Object Oriented Programming. A second key element of the ISLANDER formalism is the notion of agent's *role*. Each agent can be associated to one or more roles, and these roles define the scenes the agent can enter and the protocols it should follow. Finally, this formalism defines a graphical notation that not only allows to obtain visual representations of scenes and protocols but is also very helpful while developing the final system, as they can be seen as blueprints[16].

2.5 Formalizing the *CARREL* System

We will focus now on the formalization of the institutional aspects of *CARREL*, as it is the one that performs the mediation among hospitals and tissue banks in the transplant process.

As mentioned in the previous section, we will follow the ISLANDER formalism described in [68] to give a formal description of the *CARREL* system.[17]

2.5.1 The Performative Structure

The scenes and the connections between them constitutes the *performative structure*. This is a network of scenes that defines the possible paths that each agent role may take through the institution. In accordance with its role, an agent may or may not be permitted to follow a particular path through the performative structure, and ultimately, may be required to leave the institution.

In the *CARREL* System, the interaction among the agents and the institution is structured using the following scenes:

- *Reception Room*: is the scene where all the external agents should identify themselves in order to be assigned the roles they are authorized to play. If these agents are carrying either a request for one or more tissues or an offer of one or more organs, then this information is checked to make sure that it is well-formed.

- *Consultation Room*: is the scene where the institution is updated about any event or incident related to a piece. Agents coming from tissue banks should update the

---

[15]A comparison of different agent-based methodologies and frameworks is presented in Section 3.4.

[16]Those graphs have a similar role as UML diagrams in regular Software Engineering.

[17]In the following sections we use the original definitions and graphical notation described in [68, 188], instead of the variation introduced in the ISLANDER tool [103].

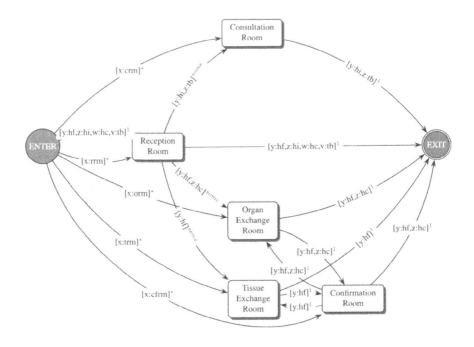

Figure 2.8: The *CARREL* performative structure

institution about tissue availability, while agents coming from hospitals should update the institution about the waiting lists and also inform it about the reception of all pieces (organs or tissues) they have received, the transplant operation and the condition of recipients.

- *Exchange Room*: is the scene where assignation of pieces takes place. In fact, there are specific exchange rooms for tissue requests (*Tissue Exchange Room*) and for organ offers (*Organ Exchange Room*).

- *Confirmation Room*: is the scene where the provisional assignments made at the exchange rooms are confirmed, whereafter a delivery plan is constructed, or cancelled, because a new request of higher priority has arrived.

A key element of the ISLANDER formalism is the definition of agent *roles*. Each agent can be associated to one or more roles, and these roles determine the scenes an agent can enter and the protocols it should follow (the *scene protocols* are defined as multi-role conversational patterns). There are two kinds of roles: the *external roles* (roles for incoming agents) and the *institutional roles* (roles for agents that carry out the management of the institution). The external roles are as follows:

- *Hospital Finder Agent* (hf): agents sent by hospitals with tissue requests or organ offers that are seen from the point of view of the institution as requests for finding an acceptable tissue or recipient, respectively.

- *Hospital Contact Agent* (hc): agents from a certain hospital that are contacted by the institution when an organ has appeared for a recipient that is on the waiting list of that hospital. The agent then enters the institution to accept the organ and to receive the delivery plan.

- *Hospital Information Agent* (hi): agents sent by hospitals to keep the *CARREL* system updated about any event related to a piece or the state of the waiting lists. They can also perform queries on the *CARREL*'s database.

- *Tissue bank notifier* (tb): agents sent by tissue banks in order to update *CARREL* about tissue availability.

The institutional roles consist of one agent to manage each scene and one agent to coordinate all the scene relationships:

- *Reception Room Manager* (rrm): manager of the *Reception Room* scene.

- *Tissue Exchange Room Manager* (trm): manager of a *Tissue Exchange Room* scene.

- *Organ Exchange Room Manager* (orm): manager of a *Organ Exchange Room* scene.

- *Confirmation Room Manager* (cfrm): manager of the *Confirmation Room* scene.

- *Consultation Room Manager* (crm): manager of the *Consultation Room* scene.

- *Institution Manager* (im): a role played by a single agent that registers all the events that happen inside *CARREL*[18] and eventually coordinates all the scene managers when the system is entering in a unsafe state. This role does not define a *centralized controller*, as we have seen in Section 1.2 that such solutions are unpractical. This role defines what Fox and Das define as a *Guardian Agent* [75]: an agent that watches the state of the system, and only acts to avoid the system entering in dangerous or unsafe states (for instance, the system is going to break a rule defined in the Spanish regulations about transplant allocation).

With all the scenes and roles identified, the performative structure can be drawn (see figure 2.8). Nodes are the scenes listed above plus *enter* and *exit* nodes in order to define beginning and ending points in the diagram. Arcs are labeled with tags of the form *[variable:role]^{access}*, where *variable* is an agent a_i, *role* is one of the identified roles for the *CARREL* system and *access* is either the number of instances of agents playing such role that can enter together into the scene, or * meaning that the agent is the one that creates the scene. The diagram in figure 2.8 shows, for instance, that scene managers go directly from the *enter* point to the scene they should manage (the * means, as mentioned before, that they are the ones creating the scene), while all the external

---

[18]Spanish law on personal data security demands that all the steps of the organ allocation process to be properly registered in logs containing all important events, in order to allow further inspection.

agents must proceed through the *Reception Room* scene in order to be registered and then be directed to the proper scene according to their roles.

The performative structure is also useful as a blueprint for the final agent architecture (see Section 2.5.5).

2.5.2   Describing the Scene Protocols

Once the performative structure is defined, the next step is to define the interactions between the agents within the scenes. We do this by means of a *scene protocol*, which defines the accepted sequences of messages that two or more agents can utter within a scene. The protocol is represented by means of a directed graph, where each node is a step or state of the conversation, and arcs are utterances. For each illocution there is an illocution scheme, which defines precisely the nature of the utterance, the roles of the sender and the receiver(s) of the utterance and the information that is exchanged. Examples of such representations can be seen in figures 2.4 to 2.5. Within the conversation graph, the grayed out nodes denote points at which agents may join the conversation, those with a double circle, where an agent may leave the conversation, while the rest are intermediate nodes.

Authentication of External Agents

As above explained, in the *Reception Room* external agents enter and are registered inside the platform. In this room an authentication mechanism based in electronic certificates ensures that external agents come only from authorized organizations (which previously received the electronic certificate to be used). Once the sender has been identified and authorized, the external agents are then directed to the proper room according to their roles.

The protocol of this scene can be seen in figure 2.4: an agent a_i makes a request for admission (1) that can be accepted (messages 3a, 3b, 3c, 3d) or refused (message 2, exit state w_1). According to the role of the incoming agent a_i:

- it is headed to the *Consultation Room* (exit state w_2),

- if it brings a request from a hospital, the request is checked (messages 4 and 5). Then agent a_i waits until the appropriate *Exchange Room* is available for the assignation (messages 6 and 7a for tissues, 6 and 7b for organs).

- if it was called by the institution to receive an organ offer, the information it brings about the recipient is checked and, if all is correct, it is then directed to the *Organ Exchange Room* that sent the call.

Registering the Recipients and the Available Pieces

In order to manage the assignation of organs and tissues, the *CARREL* institution needs up to date information about a) all the available tissues for transplantation, b) the state of

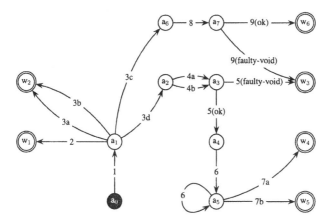

Figure 2.9: The conversation graph for the Reception Room

Msg♯	Illocution
1	(request (?x hf|hi|tb) (?y rrm) (admission ?id_agent ?role ?hospital_certificate))
2	(deny (!y rrm) (!x hf|hi|tb) (deny ?deny_reason))
3a	(accept (!y rrm) (!x hi) (accept_hi))
3b	(accept (!y rrm) (!x tb) (accept_tb))
3c	(accept (!y rrm) (!x hc) (accept_hc))
3d	(accept (!y rrm) (!x hf) (accept_hf))
4a	(inform (?x hf) (?y rrm) (petition_tissue ?id_hospital ?urgency_level ?time_to_deliver ?piece_type (?piece_parameters) (?info_recipient)))
4b	(inform (?x hf) (?y rrm) (petition_organ ?id_hospital ?time_for_availability ?piece_type (?piece_parameters) (?info_donor)))
5	(inform (!y rrm) (!x hf) (petition_state ?id_petition ok|faulty))
6	(inform (?y rrm) (?x hf) (init_exchange ?piece_type ?id_exchange_room))
7a	(request (?x hf) (?y rrm) (tissue_exchange_entrance_request !id_exchange_room))
7b	(request (?x hf) (?y rrm) (organ_exchange_entrance_request !id_exchange_room))
8	(inform (?x hc) (?y rrm) (called_for_organ ?id_hospital !id_petition)
9	(inform (!y rrm) (!x hc) (called_state !id_petition ok|faulty))

Table 2.4: The illocutions for the Reception Room

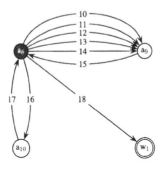

Figure 2.10: Conversation graph for the Consultation Room

Msg♯	Illocution
10	(inform (?x hi) (?y crm) (piece_arrival ?id_hospital ?id_tissue_bank ?id_piece (?state)))
11	(inform (?x hi) (?y crm) (transplantation_eval ?id_piece ?id_recipient ?date (?info_transplantation)))
12	(inform (?x tb) (?y crm) (tissue_bank_update ?id_tissue_bank ?id_piece (?specifications)))
13	(inform (?x hi) (?y crm) (waiting_list_update ?id_hospital ?id_piece ?id_recipient ?time_in (?info_recipient)))
14	(inform (?x hi) (?y crm) (maximum_urgency_level_update ?id_hospital ?id_piece ?id_recipient ?urgency_level ?time_in (?info_recipient)))
15	(inform (!y crm) (!x hi|tb) (notification_ack !id_piece ok|error))
16	(query-if (?x hi) (?y crm) (?query))
17	(inform (!y crm) (!x hi) (query_results (?results)))
18	(request (?x hi|tb) (?y im) (end))

Table 2.5: Illocutions for the Consultation Room

hospitals waiting list for each kind of organ, and c) the whereabouts about all pieces that have been assigned by *CARREL*.

The *Consultation Room* allows agents coming from hospitals or tissue banks to keep *CARREL* updated about all the facts mentioned above. The protocol of this scene is shown in figure 2.5. The incoming agents can perform notifications (messages 10 to 14) and are informed if the notification is successful (message 15). The agents coming from hospitals—which represent the Hospital Transplant Coordinator [50]—can also perform queries (message 16) about historical facts (e.g., statistics on successful cornea transplantations over a certain period). The queries are answered (message 17) with the level of detail that is permitted for a certain role, as all access to the database is controlled through a *Role-Based Access Control* model [119]. When the incoming agents have performed all the queries and notifications, they exit the *CARREL* system (message 18).

Allocating Organs

For organ assignment, a new scene, the *Organ Exchange Room* has been added. The protocol of this scene, depicted in figure 2.6, can be divided in two parts:

- the arrival of an agent a_i (hospital *Finder Agent*) with an offer of an available organ (states a_{11} and a_{12}), waiting for a notification that a proper recipient has been found (message 22, exit state w_3) or not (message 27 leading to a request for exit through state w_1).

- the loop of the scene manager looking for recipients. Based on the information of the waiting lists stored in *CARREL*'s database, the scene manager sends a call to a hospital (message 20) where there is a suitable recipient. Then an agent a_j (hospital *Contact Agent*) enters the scene to answer the call, saying whether it accepts the organ or not (message 20). Sometimes a_j, representing the hospital Transplant Coordinator, expresses the intention to use the organ in a different recipient (message 23), a change that, depending on the reasons given, can either be accepted or rejected (messages 24 and 25). If the scene manager and a_j agree, then a_i is notified of the recipient, otherwise a_j exits the scene and the loop starts again with a call to another hospital for another recipient.

The search and assignment processes performed by the scene manager are driven by knowledge of donor-recipient compatibility that is coded in the form of rules such as the following for kidneys[19]:

```
1- (age_donor <= 1)
     -> (age_recipient < 2)

2- (age_donor > 1) AND (age_donor < 4)
     -> (age_recipient < 4)
```

---

[19]It is important to note that the order the rules are presented here has no precedence meaning (for instance, rules 11 to 13 are the most important ones, and in an inference engine where rule ordering also means preference -such as PROLOG- they should be placed first). In this example the rules have been grouped by topic only for display purposes.

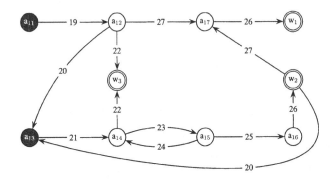

Figure 2.11: The conversation graph for the Organ Exchange Room

Msg#	**Illocution**
19	(query-if (?x hf) (?y orm) (recipient_for_organ ?id_petition))
20	(query-if (?x orm) (?y hc) (call_for_recipient ?id_recipient
	!id_petition ?time_for_availability ?piece_type (?piece_parameters)
	(?info_donor)))
21	(inform (!y hc) (!x orm) (call_answer !id_petition ?id_hospital))
22	(inform (?x orm) (?y hf) (recipient_found !id_petition !id_recipient
	!id_hospital))
23	(query-if (?x hc) (?y orm) (change_recipient (!id_previous_recipient
	?id_new_recipient ?change_reason))
24	(inform (!y orm) (!x hc) (accept_change))
25	(inform (!y orm) (!x hc) (reject_change reason))
26	(request (?x hf|hc) (y im) (exit ?exit_reason))
27	(inform (?x orm) (?y hf) (recipient_not_found reason))

Table 2.6: The illocutions for the Organ Exchange Room

```
3- (age_donor >= 4) AND (age_donor < 12)
     -> (age_recipient > 4) AND (age_recipient < 60)

4- (age_donor >= 12) AND (age_donor < 60)
     -> (age_recipient >= 12) AND (age_recipient < 60)

5- (age_donor >= 60) AND (age_donor < 74) AND (creatinine_clearance > 55 ml/min)
     -> (age_recipient >= 60) AND (transplant_type SINGLE-KIDNEY)

6- (age_donor >= 60) AND (age_donor < 74) AND (glomerulosclerosis <= 15%)
     -> (age_recipient >= 60) AND (transplant_type SINGLE-KIDNEY)

7- (age_donor >= 60) AND (glomerulosclerosis > 15%) AND (glomerulosclerosis <= 30%)
     -> (age_recipient >= 60) AND (transplant_type DUAL-KIDNEYS)

8- (weight_donor = X)
     -> (weight_recipient > X*0.8) AND
        (weight_recipient < X*1.2)

9- (disease_donor Hepatitis_B)
     -> (disease_recipient Hepatitis_B)

10-(disease_donor Hepatitis_C)
     -> (disease_recipient Hepatitis_C)

11-(disease_donor HIV)
     -> (DISCARD-DONOR)

12-(glomerulosclerosis > 30%)
     -> (DISCARD-KIDNEY)

13-(HLA_compatibility_factors < 3)
     -> (DONOR-RECIPIENT-INCOMPATIBILITY)
```

Rules 1 to 8 are related to size compatibility, either considering age ranges (rules 1 to 7) or weight differences, here the criterion permits a 20% variation above or below. Rules 5 to 7 consider *quality* of the kidney[20] and assess not only the limit that is acceptable but also the transplant technique to be used (to transplant one or both kidneys). Rules 9 and 10 are examples of diseases in the donor that do not lead to discarding the organ for transplantation, if a proper recipient is found (in the example, a recipient that has had also the same kind of hepatitis B or C in the past). Finally, rules 11 to 13 are examples of rejection rules, as determined by current medical knowledge.

It is important that such policies not be hard-coded in the system, as such rules evolve with practice (for instance, some years ago donors with any kind of Hepatitis were discarded). Expressing the knowledge in the form of rules is a technique that allows the system to be adaptable to future changes in medical practice.

Allocating Tissues

The *Tissue Exchange Room* is the place where negotiation over tissues is performed. The protocol of this scene is shown in figure 2.7:

---

[20] *Creatinine clearance* is a positive factor in kidney filtering behaviour, while *glomerulosclerosis* is a negative one.

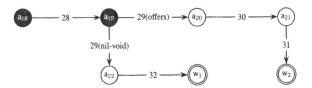

Figure 2.12: The conversation graph for the Tissue Exchange Room

Msg♯	Illocution
28	(query-if (?x hf) (?y erm) (offer_list ?id_petition))
29	(inform (!y erm) (!x hf) (offer_list !id_petition (list (?id_piece1 ?info_piece1) ... (?id_piecen ?info_piecen)))
30	(inform (?x hf) (?y erm) (weighted_list !id_petition (list (!id_piece1 ?weight) ... (!id_piece1 ?weight)))
31	(query-if (?y erm) (?x hf) (piece_offer (?id_petition ?id_piece ?cost_estimation)|void))
32	(request (?x hf) (y im) (exit ?exit_reason))

Table 2.7: The illocutions for the Tissue Exchange Room

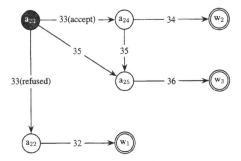

Figure 2.13: Conversation graph for the Confirmation Room

Msg♯	Illocution
32	(request (?x hf) (y im) (exit ?exit_reason)
33	(inform (?x hf) (?y cfrm) (piece_eval ?id_petition ?id_piece accept|refused))
34	(inform (?y cfrm) (?x hf) (piece_delivery ?id_petition ?id_hospital ?id_tissue_bank ?delivery_plan))
35	(inform (?y cfrm) (?x hf) (piece_reassigned_exception ?id_petition ?id_piece ?reassignment_reason))
36	(query-if (?x hf) (?y cfrm) (another_offer_list ?id_petition))

Table 2.8: Illocutions for the Confirmation Room

- a_i (hospital *Finder Agent*) asks the scene manager for tissue offers (tissues matching the requirements included in their petition).

- Then the scene manager provides a list of available tissues (message 29) that is evaluated by the external agent a_i (message 30).

- With this information the scene manager can make a provisional assignment and solve collisions (two agents interested in the same piece). When this provisional assignment is delivered (message 31) a_i exits the scene to go to the *Confirmation Room* represented by state w_2.

- There is an alternative path for the case when there are no available pieces matching the requirements described in the petition (message 9 with null list). In this case a_i requests an exit permission from the institution (message 32, exit state w_1), including the reason for leaving. The reason provided is recorded in the institution logs to form an audit trail for the relevant authorities to inspect.

For further information about this negotiation process see [48].

Confirming the Assignation

In the *Confirmation Room* scene, the provisional assignments made in a *Tissue Exchange Room* or an *Organ Exchange Room* are either confirmed or withdrawn. Figure 2.8 shows the protocol of this scene:

- The agent a_i can analyze the assigned piece data and then accept or refuse it (message 33).

- If the agent a_i accepts the piece and no higher-priority requests appear during a certain time window then the provisional assignment is confirmed and a delivery plan is given to agent a_i (message 34), and then it exits the *CARREL* system (exit state w_2).

- When there is a request with higher priority that needs the piece provisionally assigned to a_i a conflict arises. To solve the conflict the scene manager notifies the agent a_i that the assignment has been withdrawn (message 35) and that he is then entitled to a fresh request for another piece, if available, (message 36) to be negotiated again in the *Exchange Room* whence it came.

2.5.3 Normative Rules

Although we as humans perform in the context of norms in many different situations every day, we have become so used to them that we are not necessarily aware of them. The premise behind our use of electronic institutions is that we are trying to define and construct similar mechanisms to help control and simplify interaction for agents in virtual worlds. And the objectives and outcomes remain the same: to govern the behaviour of agents, establish rights and obligations and above all, to create an atmosphere of trust.

In this case, our aim is to express all the norms, from those that the national transplant organizations (such as the ONT) define to the ones defined in national and international regulations. However some of the norms are expressed at a very high level and convey no hint of how they may be implemented. For instance ONT statutes make reference to *fairness* and *equity* of organ distribution. In the case of ISLANDER, a formalism that works at the dialogue level, such norms cannot be expressed directly. Instead the designer has to *create* rules of behaviour, restrictions, constraints and, eventually, arrive to the level of individual actions, such as leaving a scene, or individual expressions, such as the protocols defined by the conversation graphs. Thus, norms can be found defined implicitly in the performative structure:

- *inter-scene constraints*: connections among scenes are, in fact, constraints about the accepted paths for each agent role inside the scene network.

- *intra-scene constraints*: the scene protocols define the accepted interactions for each agent role (what can be said, by whom, to whom and when).

The claim of the developers of ISLANDER is that all these implicit norms are made explicit by the agents while they are following the paths defined in the *performative structure* and the *scene protocols*.

But there are other norms that cannot be defined through the graphical notation. These are norms where actions taken by an agent a_i inside a certain scene may have consequences for its future interactions outside that scene. An instance of this kind of situation in *CARREL* is the arrival of a high-priority request in the Reception Room scene has effects in the Confirmation Room scene, as any provisional assignment cannot be confirmed until the high-priority request[21] has been served. The ISLANDER formalism copes with those situations by creating a list of state restrictions that should be checked regularly by the institutional agents.

For instance, the situation of the high-priority request can be expressed informally by a rule such as the following:

IF a request has arrived in the reception room,
AND the request is high-priority,
AND the request is well-formed,
AND no piece has yet been assigned to the request,
THEN there can be no assignment confirmation in the Confirmation room.

Figure 2.14 shows this rule split into its elements (pre-conditions and post-conditions) along with a formal specification of each element. The textual notation in ISLANDER of this rule is included in figure 2.15

Even though ISLANDER has the option to express those additional rules, the notation chosen does not allow the expression of other important situations such as delegation of responsibility.[22] An instance of this kind of situation in *CARREL* is the obligation of

---

[21]The so-called *urgency 0*, see Section 2.3.2.

[22]Schillo *et al.* describe in [193] two types of delegation: *task delegation*, which is the delegation of a sequence of goals to be achieved, and it has been widely studied in task-oriented approaches; *social delegation*,

`done((inform (?x hf)`	IF a request has arrived in the reception room
`(?y rrm) (petition`	
`?id_hospital ?urgency_level`	
`?time_to_deliver ?piece_type`	
`(?piece_parameters)`	
`(?info_recipient))), r-room)`	
∧	AND
`!urgency_level = urgency_0`	the request is high-priority
∧	AND
`done((inform (!y rrm) (!x hf)`	the request is well-formed
`(petition_state ?id_petition`	
`ok)), r-room)`	
∧	AND
`¬ done((inform (?z cfrm)`	no piece has yet been assigned to this request
`(!x hf) (piece_delivery`	
`!id_petition !id_hospital`	
`?id_tissue_bank`	
`?delivery_plan)), cf-room)`	
⇒	THEN
`∀ (w hf): (w ≠ x)`	there can be no assignment confirmation in
`: ¬ done(((inform`	the Confirmation room
`(!z cfrm) (!w hf)`	
`(piece_delivery ?id_petition`	
`?id_hospital ?id_tissue_bank`	
`?delivery_plan)), cf-room)`	

Figure 2.14: A norm linking actions in one scene with consequences in another

hospitals to inform the institution of all events related to the pieces that they have received from the institution. This situation can be expressed informally like this:

> IF an agent a_i is representing hospital$_h$,
> AND a_i accepts a piece$_n$ in the Confirmation Room,
> THEN another agent a_j representing hospital$_h$ must come back to the
> Consultation Room to update the database about the evolution of the
> recipient of piece$_n$

---

which consists of an agent representing another agent, a group or an organization in, e.g., a negotiation or an interaction with another parties. Hence, the *delegation of responsibility* issue we point here is a kind of *social delegation* to be addressed.

This rule *tries* to specify as closely to the action level as possible, the kind of contract that hospitals *commit* to: when accepting a piece from the institution, hospitals agree to update the institution with information. What ISLANDER fails to describe is that the contract is among two organizations, not two single agents: agents a_i and a_j may be (and in fact they surely will be in the way the UCTx has been designed) two different agents playing two different roles (a *Hospital Finder Agent* or *Hospital Contact Agent* role in the first case, a *Hospital Information Agent* in the later). At the time of writing this document no formalization of delegation is available in ISLANDER, as in all previous setups that have been modeled with it any entity or organization was represented by a single agent at any time.

A more detailed discussion about this issue is made in Section 2.6.

2.5.4   The Textual Specification

A topic of on-going work [225][39] is the use of model-checking to establish properties of the scene protocols and of the performative structure. For this reason, there is a complementary textual presentation of the illocutions, scenes, dialogic framework and performative structure, which is amenable to transformation and analysis for exactly this purpose. It must be emphasized that the textual representation is designed to capture the institutional information, rather than being directly suitable for model-checking. On the other hand, it is relatively straightforward to extract the transition networks and the arc labels from the specification and then generate textual input for model checkers such as *Spin* [97] or *nuSMV* [38]. Figure 2.15 shows part of the textual description of the *CARREL* system.[23]

2.5.5   The Agent Architecture

Once all the above framework is in place, it is possible to describe the whole multi-agent system and the relations among the agents.

It is straightforward to get the agent architecture (figure 2.16) from the Performative Structure. First of all, there is at least one staff agent for each institutional role (identified in Section 2.5.1):

- the *RR Agent* as a *reception room manager*;

- the *CR Agent* as a *consultation room manager*;

- an *ER Agent* for each exchange room, acting as an *exchange room manager*;

- the *CfR Agent* as a *confirmation room manager*; and

- the *IM Agent* as the *institution manager*, coordinating the other agents.

In order to assist the agents mentioned above, two facilitator agents are added: the *Planner Agent* which is a specialized agent with the task of with carrying out the planning

---

[23]In the last releases of the ISLANDER tool[103] this textual notation has been replaced by a new one in XML.

```
(define-performative-structure carr-performative-structure
 scenes:
  ((root root-scene)                          ; declare the
   (r-room reception-room-scene)              ; scenes comprising
   (te-room tissue-exchange-room-scene)       ; the institution
   (oe-room organ-exchange-room-scene)
   (cf-room confirmation-room-scene)
   (cs-room consultation-room-scene)
   (output output-scene))
 connections:
  ((root r-room te-room ...)))                ; and set up the connections

(define-dialogic-framework carr-dialogic-framework
 ontology: carr-ontology
 representation-language: first-order-logic
 illocutionary-particles: (......)
 external-roles: (hf, hc, hi, tb)            ; list all the roles that
 internal-roles: (im, rrm, trm, orm, cfrm, crm)  ; participating agents may play
 role-hierarchy: ((im rrm) (im trm) (im orm) ; and how the roles are
                  (im cfrm) (im crm))         ; related to one another
 role-compatibility: (incompatible: (hf tb) (hc tb) (hi tb)))

(define scene te-room                        ; describe the conversation graph
 roles: (trm hf)                             ; contained in a particular scene
 dialogic-framework: te-room-df              ; specifying which roles may
 states: (a18 a19 a20 a21 a22 w1 w2)         ; participate
 initial-state: a18                          ; the rest is just a textual
 final-states: (w1 w2)                       ; representation of a finite state
 access-states: ((trm (a18)) (hf (a18 a19))) ; machine
 exit-states: ((trm (w2)) (hf (w1 w2)))
 agents-per-role: ((trm 1 1) (hf 0 n))
 transitions:                                ; transition labels are speech acts
  ((a18 a19 (query-if (?x hf) (?y trm) (offer_list ?id_petition) ))
   (a19 a20 (inform (!y trm) (!x hf) (offer_list !id_petition
    (list (?id_piece1 ?info_piece1) ... (?id_piecen ?info_piecen)))))
   ...
   (a22 w1 (request (?x hf) (?y im) (exit ?exit_reason) )))
 constraints: ())
...

(define-norm urgency0-arrival                ; define a norm in three parts
  antecedent:
   (((r-room
      (inform (?x hf) (?y rrm)
       (petition ?id_hospital ?urgency_level ?time_to_deliver ?piece_type
        (?piece_parameters) (?info_recipient) ) ) )
     (!urgency_level = urgency_0))
    ((r-room
      (inform (!y rrm) (!x hf) (petition_state ?id_petition ok )) )
     (nil)))
  defeasible-antecedent:
   ((cf-room
     (inform (?z cfrm) (!x hf)
      (piece_delivery !id_petition !id_hospital ?id_tissue_bank
                                          ?delivery_plan))))
  consequent:
   (forall (w hf) (not (= w x))
     (obliged !z (not (inform (!z cfrm) (!w hf)
      (piece_delivery ?id_petition ?id_hospital ?id_tissue_bank
                                     ?delivery_plan) ), cf-room))))
```

Figure 2.15: Part of the *CARREL* specification in text format

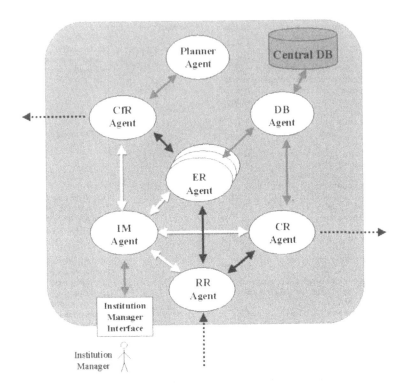

Figure 2.16: The *CARREL* agent architecture

needed in the Confirmation Room, and the *DB Agent* (an agent specialized in the access control to the Database, also depicted in the figure).

Finally, we have included an interface connected to the *Institution Manager*, to allow a human administrator to access the system.

Two prototypes of this architecture have been implemented. CARREL 1.0 was a first prototype which only mediated in tissue allocation. Later the prototype was extended: CARREL 2.0 is a version which mediates also in organ allocation. Both prototypes have been implemented in JADE [16] and they are, thereby, FIPA compliant [74].

2.5.6   A Network of *CARREL* Institutions

In the previous sections the *CARREL* system has been described as an institution that works alone, managing all the requests and offers coming from the hospitals. However a distributed system is needed in order to manage the allocation problem at an international level (one of the aims of our scheme).

To do so, we propose to create a federation of geographically-distributed *CAR-REL* platforms. Hospitals and Tissue banks register themselves to the *nearest* platform and interact as described in previous sections.

As a result, the search process becomes distributed through the platforms exchanging information among themselves via their *DB Agents*. The process is the following:

- The *DB Agent* of a certain platform$_i$ receives a query, either from an *Organ Exchange Room*, a *Tissue Exchange Room* or the *Consultation Room*

- It accesses the local database.

- If the information is not locally available, then it sends part of the query to other *DB Agents* in other *CARREL* Platforms.

- All the differences in terminology are solved at this point by the use of domain ontologies shared by all the platforms that define a common exchange format for the information.

All *CARREL* platforms are aware of the existence of the other platforms. The communication among agents on different platforms is achieved by the mechanism defined in the FIPA specification for communication among Agent platforms [74].

2.6 Discussion

In this chapter we have described the Spanish organ and tissue allocation problem and how an existing formalism such as ISLANDER can be used in the design of a model of an *e*-institution to partially automate the allocation process. Thanks to this dialogical formalism we have been able to formally specify the institution by defining the roles the agents should play in the system and then the protocols they have to follow. Once the performative structure and the scene protocols were defined, the resulting diagrams were used as blueprints to implement the first prototype multi-agent system.

We have also seen some of the problems that have arisen during such process, though. Most of them arise because the chosen approach in ISLANDER (the dialogical perspective) fails to cover all the aspect needed in complex, highly regulated scenarios such as the organ and tissue allocation we have presented:

P1 *Delegation of responsibility from agents to its owners*: For instance, as we have seen, we need to express delegation of all non-fulfilled commitments of the incoming agents to the hospitals they belong to. However, as mentioned in Section 2.5.3, there is no explicit concept of *responsibility*, and then a related *delegation mechanism*. From a theoretical point of view, one can partially simulate the delegation mechanism by means of the subsumption relation that implicitly is present inside the $\succeq$ relation that defines the role hierarchy [188], but there are two issues:

- the implementation of such relation in the execution model (also presented in [188]) does not allows delegation of responsibilities, as the only responsible of the actions an agent performs is that agent.

- when a relation such as $b \succeq a$ is established, the delegation of responsibilities from the subsumed role a to the empowered role b is unprincipled.[24] The $\succeq$ relation is a special kind of *power relation*, where one role is empowered over other role to assign responsibilities. Therefore, in the case of *CARREL*, as *hospital* $\succeq$ *hospital_finder*, delegation of responsibility should go from *hospital* to *hospital_finder* and not the other way around.

Therefore, we need more expressiveness in the formalism to handle responsibilities. In literature there are some approaches to this problem. A good example is the work of Norman and Reed [150] in terms of Deontic Logic.

P2 *Design of the protocols*: In environments such as electronic auction houses, where interaction among agents is ruled by quite simple game rules, the step of designing a protocol following the rules is not a hard one. But in environments which are highly regulated (such as the Spanish organ and tissue allocation process presented in this chapter), the design of the protocols is a very complex task: the designer has to define the behaviour of all agents in a way that follows all the regulations that are related to the problem. The designer has two possible options to do it:

- model an existing protocol for the problem: this option is based on the assumption that existing protocols, if exist, already follow all the regulations. In our case this would mean to model the organ allocation process *as it is* in the Spanish case. This solution has a main drawback: we do not use the agents in its full potential to optimize the process. For instance, the search of recipients is done by sequential calls from hospital to hospital, until a recipient is found. In the case of an agentified service, this search can be optimized by means of parallel search of possible recipients (e.g., by means of FIPA's `call-for-proposals` protocol).

- create new protocols or change current ones: this option is the usual one. As we have seen, usually optimization of process means doing some changes into the protocol, but then the problem is how to do those changes (this is the case of the adaptations we have made to the organ allocation process). Another case is when there are no pre-defined protocols to be followed (this is the case of the tissue allocation process, which now a days is not yet controlled by the ONT. As ONT aims in the future to coordinate tissue allocation, we have introduced it in our model, but we had by doing major adaptations from the protocols followed by tissue banks.

P3 *Check the protocols' compliance of regulations*: An even harder task is to formally check that the protocols follow those normatives. Following the ISLANDER formalism, Padget proposes to use *Ambient Calculus* (a process algebra with locations

---

[24] A clear example is a factory, where a *factory manager* is empowered over the *workers* of such factory ($factory\_manager \succeq worker$). It is the manager the one who assigns responsibilities, depending on the capabilities of each worker. In the case a worker is ill, the manager cannot assume its responsibilities automatically, as usually the manager has no qualification or skills to operate the machinery of the factory.

[30]) in order to obtain a precise description of the interaction of the institutional agents with the external agents [162]. However:

- no formal link is presented between the ambient calculus formalism, the graphical notation and the list of additional rules in order to do a formal model checking, and

- there is no mechanism to relate those protocols with the regulations that should be followed in complex environments.

The latter issue (the relation among norms and protocols) is quite important in the field of *e*-institutions, as one important part of the definition of *e*-institution (that we presented in Section 1.3.3) is the concept of norms as the mechanisms to rule the behaviour of the agents. In an environment such as human transplants, that is highly regulated, norms are therefore coming not only from the institution itself but also from the environment, where some of the regulations and procedures are defined and may even be updated.

With all these issues in mind, it is now clear that we need to find a more powerful approach that can work not only in a low level of abstraction (protocols, procedures) but also cope with elements that are defined in higher levels of abstraction (norms, rules). To do so we should find a proper definition of all the concepts involved in order to create a solid foundation for a new framework. In Chapter 3 we present a review on some of the research on norms made, not only in Computer science but also in Sociology and Legal theory. Then in Chapter 4 and Chapter 5 we will define a new framework, that we will then apply to the organ and tissue allocation problem, addressing the problems (P1, P2, P3) mentioned above (Chapter 6).

To end this discussion we want to note that, despite all the drawbacks mentioned here, ISLANDER has been a very valuable contribution to the field as it was the first framework to formalize electronic institutions, and it can be used in scenarios with simple, static rules such as electronic auction houses. But for our purposes we need more powerful tools in order to model a complex domain like the one we have chosen.

Chapter 3

Norms, Agents and Electronic Institutions

Most of the descriptions of institutions are based in the formal definition of their *norms* and their effect in the actions performed by the members (i.e., the entities that belong to the institution and which follow its norms). With this idea in mind, in this chapter we will analyse the concept of *Norm*. To do so, we will analyse the view several branches of science –such as *sociology*, *economics* or *legal theory*– have of it. We will also analyse how they are modelled in *computer science* and how they are applied to Multi-Agent systems by the creation of *Electronic Institutions*.

3.1 What is a Norm?

It is hard to find a definition of *norm* that summarizes the different approaches of the norm concept in all the related scientific disciplines, from social philosophy, psychology and sociology (whose roots in the western tradition are in ancient Greek philosophers such as Socrates, Plato and Aristotle), to legal theory (whose roots come from Roman Law) and computer science.

According to sociology, a *norm* is a rule or standard of behavior shared by members of a social group (Encyclopedia Britannica). According to philosophy, a norm is an authoritative rule or standard by which something is judged and, on that basis, approved or disapproved (Columbia Encyclopedia). Examples of norms include standards of right and wrong, beauty and ugliness, and truth and falsehood. According to economics, a norm (from *norma*, Latin for carpenter's level) is a model of what should exist or be followed, or an average of what currently does exist in some context, such as an average salary among members of a large group.

In the following sections we will present the view of norms in *Social Theory*, *Legal Theory* and *Computer Science*.

3.1.1 The View of Norms in Social Theory

The study of formal and informal frameworks to model the regulations that shape interactions among humans are not a brand new field, but an ancient concern that can be found in ancient Greece and Rome. Plato's moral philosophy (which can be found in his *Republic*) is a good example. Plato stated that moral thinking must be integrated with our emotions and appetites in order to live in a society. Aristotle extended Plato's moral philosophy to the study of *ethics*, being the first one known to write ethical treatises.

Sociology is a discipline that results from an evolution of moral and ethical philosophy in order to describe the interactions that arise among the members of a group, and the social structures that are established. One of the main concepts we find in complex social structures is *role*. A role is a description of the tasks and objectives to be performed by an entity. The idea is that it is not important who plays the role as far as there are enough entities enacting it. Roles have been extensively studied in the Organizational Theory field, in order to study the relationships among the social roles an individual may play, the obligations and authorizations that are associated to each one of those roles, and the interaction of roles in the distribution of labour mechanisms.

While classical approaches used to study a society or an organization as a closed system where all the relations among individuals can be defined and established, modern approaches are focused on an open systems view where the environment or context has an important influence as it *constraints*, *shapes*, *penetrates* and *renews* the society or organization [195]. In this new scenario problems such as consensus or limited trust may affect the interactions of the individuals when trying to achieve coordination or cooperation. Human societies have successfully coped with similar problems of *social order*, mainly by the development of *norms* and *conventions* as behaviour specifications to be followed by individuals.

A good example of the view of norms in Sociology is the work of Tuomela in [206]. Tuomela distinguishes two kinds of social norms:

- *r-norms*: rules created by an authority structure and are based on agreement-making. They can be either *informal*, connected to informal sanctions, or *formal*, connected to formally defined sanctions.

- *s-norms*: social norms based on mutual belief (*conventions*). There are s-norms applied to a specific group (e.g., a group of friends), to an organization or to the whole or part of a human society.

The study of the role of norms in open systems by sociologists and economists gave birth to the so called Institutional Theory [151][196]. It is based in the fact that, in some societies, norms are supported by social institutions, which enforce the fulfillment of the norms by the members of the society. Research is mainly focused on two approaches: a) the study of the dynamics of the regulatory framework (how norms emerge, how they are imposed, how they are spread), and b) the effects of regulatory frameworks in the dynamics of a social system and in the behaviour of its individuals.

For instance, North [151] has studied the effect of normative systems (that he calls *institutions*), on the behavior of human organizations. Although North includes human

	Regulative pillar	**Normative pillar**	**Cultural-Cognitive pillar**
Basis of compliance	Expedience	Social Obligation	Taken-for-grantedness Shared understanding
Mechanisms	Coercive	Normative	Mimetic
Elements	Rules, Laws, Sanctions	Values, Expectations	Common beliefs, Shared logics of action, Categories, Typifications
Basis of legitimacy	Legally sanctioned	Morally governed	Comprehensible, Recognizable, Culturally supported
Routines	Protocols, Standard operating procedures	Jobs, Roles, Obedience to duty	Scripts

Table 3.1: Scott's *Three Pillars of Institutions*.

societies in his analysis, his studies have been mainly focused on the performance of organizations. He concludes that institutional constraints ease human interaction, reducing the cost of this interaction by 1) ensuring *trust* between parties, and 2) shaping choices. Once the institutional constraints are established, individuals are able to behave –and expect others to behave– according to the norms.[1]

An in-depth review of the research in Institutional Theory is presented by Scott in [196]. As a result of this review, Scott defines a unifying framework to describe the interactions between institutions and organizations. In such framework Scott states the following distinction between *norms* and *rules*:

- **rules** are meant to describe *regulative aspects* which constrain and regularize behaviour,

- **norms** are meant to describe *prescriptive, evaluative and obligatory aspects* of social life.

Scott's work is complementary to North's, as he focuses in the social aspects of institutions. His framework is summarized in what he calls the *Three Pillars of Institutions* (see table 3.1), where he states that an institution is composed by the regulative aspects, the normative ones and the cultural-cognitive ones (shared conceptions, common beliefs). In Scott's conception, rules are related to legal (coercive) aspects, while norms are related to moral aspects.

It is of special interest for our work the relation Scott defines between *roles* and *norms*:

---

[1] We already presented North's approach in Section 1.3.

Some values and norms are applicable to all members of the collectivity; others apply only to selected types of actors or positions. The latter gives rise to roles: conceptions of appropriate goals and activities for particular individuals or specified social positions. These beliefs are not simply anticipations or predictions but prescriptions -normative expectations- of how the specified actors are supposed to behave.

Therefore, roles are highly coupled with norms in a normative system,[2] and can be used to confer rights and responsibilities to individuals (the actors enacting the role).

3.1.2 The View of Norms in Legal Theory

The creation and use of norms is an activity that has evolved in human societies from centuries in order to ensure fairness and trust in human interactions. In most primitive cultures there were no written laws (*informal norms*) but shared moral and conventions that were followed by the individuals of the society. In order to solve conflicts, sometimes people requested the help of wise men that had a better understanding of the conventions.

Some cultures created a corpora of written laws (*formal norms*) in order to explicitly express the morals and conventions.[3] One good example was the Roman civilization, which created normative systems based on the idea of Law as an interpretation of *codices* where all values and norms are expressed. This view is the origin of the *Roman Law*, later extended by Germans, which fully formalized the different aspects of a legal system. Nowadays most of the European countries that were part of the Roman Empire (such as Italy, France, Spain, Portugal, Germany or Greece), the ones that were linked to the first by monarchy marriages (such as Austria or the Netherlands) and the Spanish, Portuguese and French colonies (such as Latin America and some old colonies in Africa) have legal systems based in Roman-Germanic Law. In most of them the normative system is hierarchical:

- the most important norms are stated in a written *constitution*,

- then the norms stated in *laws* are next in the hierarchy,

- finally the *regulations* are the less important ones.

This three-layered approach allows not only to assess if a given situation is *legal* or *illegal*, but even the legality of the corpora of norms and rules themselves. Therefore:

- regulations can be reviewed and then stated to be *illegal* if they break one or more laws, or

- regulations and laws can be considered *non-constitutional* if they break one or more of the articles in the constitution.

---

[2] We present our view of roles and their relation with norms in Section 4.2.5 and Section 5.7.

[3] A well-known example is the code of Hammurabi (1792-1750 bC.), that is a collection of 282 case laws including economic provisions, family law (marriage and divorce), as well as criminal (assault, theft) and civil law (slavery, debt).

A different example of written law is the *Common Law*, where law is composed by an aggregation of past decisions that are written for future use.[4] The United Kingdom is one of those countries with a law system based in *Common Law*, where there are no written constitution or a hierarchical system of laws and rules. In the United Kingdom the law is based in a case-based approach (a decision on a current issue is based on searching for a similar decision taken in the past). The United Kingdom's most important figure in the Normative System is the *Parliament*, which is the one that may decide if something is *legal* or *illegal*, either by past similar decisions that the Parliament made or by taking decisions in a new topic, decisions that are registered for future use. In the United States of America, although the law system is also based in the Common Law, there is a written Constitution as a backbone of the system. Therefore, situations or actions can be stated as *legal* or *illegal* depending on past decisions or based in the Constitution. And even past decisions of a governor or a judge can be revisited and stated as *legal* or *illegal* depending on if they are *constitutional* or *non-constitutional*.

We will focus on Roman Law[5], as there is already an important amount of work done on formalization of this kind of legal systems.[6] Roman Law defines two kinds of constraint sets:

- *Normatives or Laws*: set of *norms* that define WHAT can be done by WHO and WHEN. A constitution is the highest law in a state.

- *Regulations*: set of *rules* that expand a given normative defining HOW it may be applied.

One of the main problems in legal theory is that norms and rules are always expressed in natural language, so they are open to interpretation. In 1926, Ernst Mally proposed the first system of *Deontic Logic* [123]. He aimed to lay the foundation of "an exact system of pure ethics", attempting to formalise normative and legal reasoning. Mally based his formal system on the classical propositional calculus formulated in the *Principia Mathematica*. Two decades later Von Wright presented in [219] a formalism (based on *Propositional Calculus*). But then it was realized that Von Wright's formalism was very close to a normal modal logic, enabling a clear *Kripke semantics* using O as the basic *necessity operator*. The reformulation of Von Wright's system as a modal logic is called the *standard system of deontic logic* or KD, and includes the following axioms:

$$O(p \to q) \to (O(p) \to O(q)) \quad \text{(KD1 or K-axiom)}$$
$$O(p) \to P(p) \quad \text{(KD2 or D-axiom)}$$
$$P(p) \equiv \neg O(\neg p) \quad \text{(KD3)}$$
$$F(p) \equiv \neg P(p) \quad \text{(KD4)}$$
$$p, p \to q \vdash q \quad \text{(KD5 or Modus Ponens)}$$
$$p \vdash O(p) \quad \text{(KD6 or O-necessitation)}$$

---

[4]Researchers in AI & Law also refer to *Common Law* approaches as *Precedence-Based Systems*.

[5]From now on, we will talk about *Roman Law* or *Rule-Based Systems* to refer to the *Roman-Germanic Law* approach.

[6]While research in *Rule-Based Systems* is mainly focused on formal methods, research in *Precedence-Based Systems* focuses on less formal, applied approaches, mainly in the design of architectures for implementation.

where O, P and F are modal operators for *obligation*, *permission* and *prohibition*. Some authors also add a rule $KD0$ to include some tautologies from Propositional Calculus.

In Deontic Logic, a norm is viewed as an expression of the obligations and rights an entity has towards another entity, or a society. Therefore, Deontic Logic has its origins in *Moral Philosophy* and the *Philosophy of Law*. The main assumption is that verbs such as *should* or *ought* can be modelled as modalities (e.g., *"it should happen p"* , *"it ought to be p"* or *"α should be done"*). The Kripke semantics of the O, P and F operators defines, in terms of possible worlds, which (possible) situations are *ideal* according to a given normative system.

It is important to note that, although Deontic Logic is one of the most used logics to formalize normative systems, it is not a logic of norms (it is unclear if norms have truth values to be assigned), but a logic of norm propositions (i.e., propositions stating the existence of norms) [137].

3.1.3 The View of Norms in Computer Science

Computer Science is another field which has been attracted to the study of norms, specifically in the agent community, in order to solve the coordination issues raising in open multi-agent systems. In the last years, Multi-Agent Systems have become larger and larger distributed systems, where information and tasks are distributed among several *agents* in order to be able to model complex, dynamic domains such as *e-commerce*. This interest has lead to a blooming in the field, resulting in several views of what a norm is, how norms may be modelled and how agents should *follow* those norms. From the society of agents point of view, new concepts such as virtual organizations and agent-mediated electronic institutions have emerged.

The origins of this research comes from the *decision theory* field. For instance, lots of research has been made in game scenarios to identify and model the rules in *game theory*, and apply them to agent-mediated systems in environments such as the *e*-commerce.

As explained in Chapter 1, the design and implementation of large multi-agent systems is a complex task, as it is not easy to translate the requirements of such system (the collective behaviour) in a set of individual behaviours. Developers should ensure that all agents will understand the utterances of the others and will perform as expected (e.g., if an agent a_i asks agent a_j to perform a given action A, a_j certainly will do it). Another issue is to ensure that the collective behavior that will emerge from the composition of the individual behaviours is the desired one.

In the case of open systems, it is mandatory to define new mechanisms to ensure trust in complex environments. There are several research groups that have searched for analogies in human societies to find solutions that can be used in the agent societies. As explained in Section 3.1.1, one way to inspire trust in the parties is by incorporating a number of regulations (*norms*) in the organization that indicate the type of behavior to which each of the parties in the transaction should adhere within that organization. Norms are defined inside *institutions*. Hence, Institutions are established to regulate the interactions among agents that are performing some (business) transaction.

In the case of computer science, research is focused in two areas:

- how to fully formalize human normative systems in a computational representation. Research in this area focuses on a) expressiveness of logic representations [172][94][170][171] in order to model complex issues that appear in human normative systems (such as defeasibility or exceptions to a norm), and b) ontological representation of all the knowledge needed in legal reasoning (such as the CLIME [22] and POWER [211] ontologies), including dynamic aspects of normative systems such as *amendments, extensions, derogations* or *substitutions* [79];

- how the norms can be incorporated in the structure of an agent-mediated organization such that the agents operating within the organization will operate according to these norms or can be punished when they *violate* the norms.

In the following sections we will focus on the latter, where multi-agent systems are seen as *social systems*, as it is the most relevant for our work.

In each social system one can study the social interaction between the parties in the system from two different viewpoints: from the *individual's point of view* or from the *society's point of view*. The first one is mainly interested in the way an individual handles the interaction:

- How does it react to signals from other parties?

- Does it have a fixed protocol to handle interactions or does it reason about the signals?

- What is the *best* way to react to a signal?

From the society's point of view, interest is mainly focused on the properties of the interaction itself and the behaviour of the whole organization:

- Is the interaction *fair*?

- Is the interaction *efficient*?

- Are the objectives of the organization *met*?

Although the two viewpoints deal with different concerns, they are of course not completely independent. If an interaction protocol that is defined from the society's point of view requires a party to perform some action in order to respond then, of course, one should make sure that all parties are capable to perform this action. For example, if a request for information should be answered with the information (if available) in some (belief) database then the parties should be able to access their database.

3.2 Social Interaction from the Individual's Point of View

Within this view, research focuses on the study of norms and their effects in agents' behaviour. To do so there is some work on norm formalization, i.e., defining logics expressive enough to model *accepted behaviour* through norms (rights, authorizations, etc.), while other researchers focus on the study of how norms can affect the agent's behaviour, and more concretely, how they can be introduced inside the reasoning cycle.

3.2.1   Formalizing Norms

As Tuomela explains in [206], in human societies norms can be either *informal*, when there is no formal regulation of the norm and no formal sanction for violating the norm, or they can be *formal*, when they are incorporated in laws or regulations from the institutions that regulate the behavior of the people within the society. In this section we will review the work on formal norms (that are the ones that are explicitly represented in institutions). See [45, 47] for more information on the treatment of informal norms.

All the logic formalizations of norms are based in the *obligation* concept. One of the first approaches to the introduction of obligations in MAS can be found in Shoham's Agent-Oriented Programming (AOP) [198]. In this approach the obligations are expressed as *restrictions* which ensure that agents perform in a certain way.

However, there are other experts that are looking for ways to make obligations to *guide* the behaviour of the agents instead of *restricting* it. Taking the *Agents' Theory* as starting point, they have developed concepts and methodologies which can be applied to the reasoning about obligations. A good example is the work of Conte and Castelfranchi in [46], which proposes an extension of Cohen and Levesque's *"Theory of Rational Action"*[7], extending the $BEL, GOAL, HAPPENS$ and $DONE$ modal operators with a new operator $OUGHT$, in order to include obligation as another operator inside the formalism. The resulting formalism allows them to express concepts such as *normative belief* $(N-BEL)$, *normative belief of pertinence* $(P-N-BEL)$ and *normative goal* $(N-GOAL)$ as follows:

$$
\begin{aligned}
(N-BEL \; x \; y_i \; a) &= (\Lambda_{i=1,n} \, (BEL \; x \; (OUGHT \; (DOES \; y_i \; a))))\\
(P-N-BEL \; x \; a) &= (\Lambda_{i=1,n} \, (N-BEL \; x \; y_i \; a)) \wedge (V_{k=1,n} \, (BEL \; x \; (x = y_k)))\\
(N-GOAL \; x \; a) &= (R-GOAL \; x \; (DOES \; x \; a)(P-N-BEL \; x \; a))
\end{aligned}
$$

where x and y_i are agents, and a is an action. Although their approach is well suited for the study of the effect of norms in the agent goals, it lacks some expressiveness when modelling the norms in complex normative systems.

An alternative is the use of Deontic Logic to model norms. [139]. Standard Deontic Logic (the KD system) is quite expressive to analyse how obligations follow one from the other and also to find possible paradoxes in the (deontic) reasoning. However, the KD system hardly can be used from a more operational approach in order to, e.g., decide which is the next action to be performed, as it has no operational semantics. One interesting extension of KD Deontic Logic is the Dyadic Deontic Logic, proposed by von Wright, which introduces *conditional obligations* with expressions such as $O(p|q)$ (*"p is obligatory when condition q holds"*). Some of the properties are the following:

$$
\begin{aligned}
O(p \wedge q|r) &\equiv O(p|r) \wedge O(q|r)\\
O(p|q \vee r) &\equiv O(p|q) \wedge O(p|r)\\
\neg(O(p|q) &\wedge O(\neg p|q))\\
P(p|q) &\equiv \neg O(\neg p|q)
\end{aligned}
$$

---

[7]The theory of Rational Action [40] introduces a Modal Logic about Desires, Beliefs and Actions.

Conditional deontic expressions are useful to introduce weak cause-effect and temporal relations. For the latter, there are also specific logics to address temporal aspects, as the Temporal Deontic Logic. For instance,

$$O(p < q)$$

states that *"p is obligatory before condition q holds"*. Another option is combining deontic operators with Dynamic Logic:

$$[p]O(q)$$

means *"after p is performed it is obligatory q"*.

Some researchers have enhanced deontic logic by adding action modal operators E, G, H. Operator E comes from Kanger-Lindahl-Pörn logical theory, and allows to express direct and successful operations: $E_i A$ means that an agent i *brings it about* that A (i.e., agent i makes A to happen and is directly involved in such achievement). Operators G and H were introduced later by Santos, Jones and Carmo to model indirect actions [190][191]. Thus, the modal operator G allows to express indirect but successful operations: $G_i A$ means that an agent i *ensures* that A (but not necessarily is involved in such achievement). The operator H expresses attempted (and not necessarily successful) operations: $H_i A$ means that an agent i *attempts to make it the case* that A. Some of the main axioms of the E, G, H logic are the following:

$$E_i A \rightarrow A$$
$$(E_i A \wedge E_i B) \rightarrow E_i(A \wedge B)$$
$$E_i A \rightarrow G_i A$$
$$G_i A \rightarrow H_i A$$
$$E_i E_j A \rightarrow \neg E_i A$$
$$G_i G_j A \rightarrow G_i A$$
$$G_i A \rightarrow G_i E_i A$$

The advantage of having such modal action operators is that they can be combined with deontic modal operators to express more complex statements such as:

$$E_i O G j A$$

meaning that agent i is directly involved in bringing about the obligation that agent j should ensure A. With this kind of expressions it can be modelled the idea of *normative influence*, as we will see in Section 3.3.1. As Gobernatori *et al.* pointed out in [88], the H operator becomes very useful in normative domains in order to express weak normative influence without using the O deontic operator, by means of the following axiom:

$$E_i O G_j A \rightarrow H_i G_j A$$

expressing that if agent i brings about the obligation that agent j should ensure A, then agent i attempts that agent j ensures A.

Limitations of Norm Formalization

As explained in Section 3.1.2, deontic logic is not a logic of norms but a logic of propositions about norms. Although it is possible to capture the norms in this way, to give them a certain kind of semantics and to reason about the consequences of the norms, this kind of formalization does not yet indicate *how* the norm should be interpreted within a certain context. For instance, we can formalize a norm like *"it is forbidden to discriminate on the basis of age"* in deontic logic as

$$F(discriminate(x, y, age))$$

stating that *"it is forbidden to discriminate between x and y on the basis of age"*. However, the semantics of this formula will get down to something like that the action `discriminate(x,y,age)` should not occur. It is very unlikely that the agents operating within the institution will explicitly have such an action available, though. The action actually states something far more abstract.

3.2.2   Norm Autonomous Agents

In Section 1.1 we presented a Verhagen's definition of *Norm Autonomous Agent*, as agents that are capable to choose which goals are legitimate to pursue, based on a given set of norms. A more accurate definition is given by Castelfranchi *et al.* in [36]: a *Norm Autonomous Agent* (also called *Deliberative Normative Agent*) is an agent:

- able *to know that a norm exists in the society* and that it is not simply a diffuse habit, or a personal request, command or expectation of one or more agents;

- able *to adopt this norm* impinging on its own decisions and behaviour, and then

- able *to deliberatively follow* that norm in the agent's behaviour, but also

- able *to deliberatively violate* a norm in case of conflicts with other norms or, for example, with more important personal goals; of course, such an agent can also accidentally violate a norm (either because it ignores or does not recognize it, or because its behaviour does not correspond to its intentions).

Whenever norms are used within the context of multi-agent societies and institutions, the question that arises is how the agents should cope with these norms. Billari explains in [19] that norms can be introduced into the agents' decision making as either:

a) *restrictions*: in this case norms are reduced to constraints forcing compliance of norms.

b) *goals*: in this case norms are incorporated as (prioritized) goals to be included in the reasoning cycle of the agent.

c) *obligations*: norms are explicitly represented as another element of the agents reasoning cycle.

In options a) and b) norms are hard-coded in the agents through action restrictions or a restricted number of (accepted) goals, so agents will always behave according to these restrictions In this scenario agents do not have to *know* that these restrictions follow from specific norms. Therefore, the overhead in reasoning about the possible consequences of their behavior is (almost) zero, allowing the creation of relatively simple and small agents. In option c) norms are explicitly represented through obligations, so agents can be designed to reason about the norms. These agents might exhibit the same behavior in most circumstances as the simple agents, but will be able to make better choices in special circumstances. They know when it pays off to violate a rule. As these agents have to reason about the possible consequences of every action, they are quite more complex.

F. Dignum further explains in [55] how agents reason about violating a norm. It should be noted that the proposal states that norms should not be imposed, they should only serve as a guidance of behaviour for agents, or agents will loose their autonomy.

With this idea in mind, F. Dignum presents in [55] an agent architecture where social norms are incorporated to the BDI cycle of the agent. In that architecture, norms are explicitly expresed in deontic logic by means of the obligation operator O and are divided in three levels:

- *convention level*: norms modelling the social conventions.

- *contract level*: norms regarding the obligations that arise when an agent a_i commits to either perform an action α or to achieve a situation p requested by agent a_j.

- *private level*: the intentions of the agent are seen as commitments towards itself to perform a certain action or plan.

At the convention level, the general social norms and values are defined in terms of obligations of any agent a_i towards the whole society and are expressed in deontic logic as $O_i(x)$. At the contract level the commitment between two entities a_i and a_j is modelled as an obligation of a_i towards a_j, which is then expressed in deontic logic as $O_{ij}(\alpha)$ or $O_{ij}(p)$. Formally, the link between a commitment and an obligation can be expresed as follows:

$$[COMMIT(i, j, deliver)] \, O_{ij}(deliver)$$

In the formalism there is also a representation of *authorizations* that agents can give to other agents. An example is the following:

$$auth(j, ASK(i, j, pay)) \rightarrow [ASK(i, j, pay)] \, O_{ji}(pay)$$

This formula means that agent j authorizes agent i to ask him for a payment, and this is translated as an obligation of j towards i to pay.

To assure completeness of the model, the BDI interpreter presented by Rao and Georgeff [177] is extended to work with obligations. At this level, the intentions of the agent are seen as commitments towards itself to perform a certain action or plan, and are modelled as obligations of the form $O_{ii}(\alpha)$. The decision making process is then driven

by a list of obligations, and a preference relation is defined to choose among them the one
to be fulfilled next.

The main drawback of Dignum's approach is that is too agent centered, and it does
not explain how norms, at the convention level, are organized and how to model an entire
organization by means of those conventions.

3.2.3 Discussion about Agent-Centric Views

In summary, we have seen that those approaches which study norms from the agents'
perspective are mainly focused on the effect of norms in the agents beliefs, goals and plans
(sequences of actions). Some approaches study this effect of norms from a theoretical
perspective, modelling norms by means of formalisms more or less connected with the
Deontic Logic, while others extend previous agent theories to analyze the relationship
between intentions and obligations, that is, how agents decide to fulfill or to violate an
obligation depending on their internal beliefs and goals.

Although it may seem that agent-centric approaches are well suited to study how
agents should behave in open environments –as they model the effect of normative frameworks in the agents' reasoning cycle–, there are two drawbacks:

- agent-centric views lack a good model of the society that surrounds the agent, that
  is, a model of the (pre-stablished or emerging) social structure that describes the
  relations of the agent with other agents;

- in [224], Wooldridge and Ciancarini describe a problem of those formalisms and
  agent theories based in *possible worlds* semantics. They state that there is usually
  no precise connection between the abstract accessibility relations that are used to
  characterize an agent's state and any concrete computational model. This makes it
  difficult to go from a formal specification to an implementation in a computational
  system.

In order to address Wooldridge and Ciancarini's issue, Lomuscio and Sergot present
in [124] the *deontic interpreted systems*, a formalism that tries to capture some of the
ideas of deontic reasoning through a more operational approach. It defines two new modal
operators, $\mathcal{O}$ and $\mathcal{P}$. $\mathcal{O}_i\alpha$ stands for *"in all the possible correctly functioning alternatives
of agent i, α is the case"*, where the concept of *correctly functioning alternatives* makes
reference to the compliance of a given protocol the agent i has to follow.[8] The $\mathcal{P}$ operator
is the dual of $\mathcal{O}$: $\mathcal{P}_i\alpha$ stands for *"in some of the states in which agent i operates correctly,
α holds"*. The aim of these operators is to represent local and global states of violation
(*red states*) and compliance (*green states*) with respect to some defined protocols. Although Lomuscio and Sergot also present an axiomatization of their *deontic interpreted
systems*, their formalism fails to model rich relations among agents (the social structure)
by means of, e.g., *roles*, as their approach is also agent-centric.

---

[8]The $\mathcal{O}$ operator in Lomuscio and Sergot's formalism is related with the O operator in Deontic Logic but
they are not equivalent: while $O_i\alpha$ expresses that *"it is obligatory for i that α happens"*, $\mathcal{O}_i\alpha$ only expresses
the operational view that α will happen in the future if the agent follows the correctly functioning alternatives
defined in a protocol.

3.3 Social Interaction from the Society's Point of View

Within this view, research is focused on the study of different issues regarding social interaction and social order, always from the society (or organization) point of view. Hence, groups of agents that are somehow connected (e.g., because of continuous interaction, shared resources) are seen as *agent societies*. There are two main branches in the research:

- *Definition and formalization of Social Systems*: this area of research focuses on the creation of formal and non-formal models to describe societies or organizations, their members, the social structure and, in some cases, the normative framework that shapes the behaviour of the individuals. The foundations of this area are mainly *sociology* and *organizational theory*;

- *Simulation of Social Systems*: there has been quite a lot of work in simulating different aspects of social dynamics, like division of labour or, more related to our work, norm emergence and norm acceptance. The foundations of this area are mainly *evolutionary theory*.

In this section we will focus on the former, as the norm formalisms they propose are better suited for our objectives. In the case of the latter, in order to run simulations about *norm spreading* and *norm internalization* within a society, norms are reduced to *filters* (an example is the work of Egashira and Hashimoto [65]) or *decision trees* (an example is the work by Verhagen [216]), which may be valid for simulation but not for a proper modelling of a normative system.

3.3.1 Definition and Formalization of Social Systems

In literature there is a variety of models, languages, frameworks and methodologies which aim to extend the agent-centric approaches with concepts such as *global* or *social goals*, *social structures* and *social laws*. In this section we will present some examples of the different approaches to model social order, namely the *game-theoretic* approaches, the *sociological* ones, the *organizational* ones and the *logic* ones. We will focus in those approaches that include in their definition some kind of normative system.

A Game-Theory Approach to Social Order

Game theory has been one of the most used approaches in Multi-agent systems to model different types of interaction between agents (e.g., *cooperation*, *competition* or *negotiation*). However, in most of the research such interaction is governed by sets of *game rules* that are not expressive enough to model complex normative systems.

An interesting exception is the work by Shoham, Tennenholtz and Moses. In [144, 199], they present a computable mathematical formalization of agent societies that includes a *normative level*. They define the concept of *Artificial Social System* as a set of restrictions on agents' behaviours in a multi-agent environment. An Artificial Social System is indeed a mechanism for coordination that reduces the need for centralized control, allowing agents to coexist in a shared environment and pursue their respective goals in the presence of other agents.

The formalization they propose is an automata-based model of multi-agent activity that describes the system as a set of agents that, at any given point in time, are each of them in one state. The definition of such system of dependant automata (DA system) S is the following:

- L_i denotes the set of local states of an agent a_i.

- the DA's *system configuration* is expressed by the list $\langle s_1, \ldots, s_n \rangle$ of states of the different agents in the DA

- for every state $s \in L_i$, there is a set $A_i(s)$ of actions that i is capable to perform when it is in local state s.

- a *plan* for agent a_i in a DA system is a function $p(s)$ that associates with every state s of a_i a particular action $a \in A_i(s)$

With this first definition there are no restrictions of the actions each agent can perform in the system. Therefore, computing plans in a way each agent will achieve its goals regardless of what the other agents do is a very complex task, and may be even impossible. In order to guarantee that the agents may be able to succeed in fulfilling at least some of its goals, the DA system is modified in order to introduce *social laws* that restrict the set of actions an agent is *allowed* to perform at any given state:

- a social law Σ consists of functions $\langle A'_1, A'_2 \ldots, A'_n \rangle$, satisfying $A'_i(s) \subseteq A_i(s)$ for every agent a_i and state $s \in L_i$.

- S^Σ denotes the new DA system resulting from changing in S the functions A_i by the restricted functions $A'i$.

Intuitively, the last point means that in S^Σ the agents can behave only in a way that is compatible with the social laws. In [144, 73] some criteria for designing social laws are proposed. The foundation of these criteria came from a game-theory approach where some kind of equilibrium among individual plans (which are seen as strategies) is sought:

- G_{soc} is the set of *socially acceptable goals* for each agent. They determine which strategies are *legal* and which are not.

- S is the set of possible strategies, identical for all agents.

- for each goal $g \in G_{soc}$ a payoff function $u_g : S^n \longrightarrow [0, 1]$ is associated. Also an *efficiency parameter* $0 \leq \varepsilon \leq 1$ is defined.

- $\bar{S} \subseteq S$ is the set of socially acceptable strategies such that for all $g \in G_{soc}$ there exists $S \in \bar{S}$ such that $u_g(s, \sigma) \geq \varepsilon$ for all $\sigma \in S^{n-1}$.

While designing a social system, the designer needs to find a good compromise between competing objectives. On one hand, it should disallow some of the possible strategies in order to ensure efficient achievement of certain goals, while on the other hand it is necessary to maintain enough strategies in order that agents can attain their goals in a reasonably efficient manner.[9]

In order to provide the *Artificial Social System* formalism with a clear semantics, they use the concept of *possible worlds*. Their formalization can be summarized as follows:

- W is the set of *possible worlds*,

- $K_i \subseteq W \times W$ are *accessible relations*,

- A is a set of *primitive individual actions*,

- $Able_i : W \longrightarrow 2^A$ is a function determining the *possible physical actions* for agent a_i in any world W.

- for all agents, the set $Legal_i \subseteq Able_i$ defines the *legal system*. It is required that for all agents and all worlds, $Legal_i(w) \neq \emptyset$.

- W_{soc} is the set of *socially acceptable worlds*.

- A world w is *legally reachable* only if $w \in W_{soc}$.

Last two points capture the idea of a *social system* to ensure that agents will be able to attain any *reasonable* or *socially acceptable goal*[10]. To do so they add an important restriction: the social system should guarantee that the state of the world will never exit the set W_{soc}. In a sense, the set W_{soc} is the one that captures the global aspects of the behaviour of the system (mainly the *safety* and *fairness*).

Even though this formalization captures the idea of normative systems as a set of socially acceptable goals and states and it has been used in some setups such as mobile robots [199, 154], it has some drawbacks:

- The restrictions are applied to individual agents. There is no definition of roles.

- one of the assumptions of the model is that all agents have the same set S of strategies and the same payoff function (which depends only on its current goal and the strategies executed). So the behaviour of any agent is foreseeable at any time given its current goal and state.[11]

---

[9]Moses and Tennenholtz call this compromise the *Basic Golden Mean Problem*.

[10]One interesting statement of Moses and Tennenholtz in [144] is that a *normative system* does not ensure *fairness*. They provide the following example: There are several agents that want to eat a piece of cake. The designer may try to ensure fairness by placing the restriction that an agent cannot eat all the cake (leaving nothing to the others), but the action of two or three agents may result in other agents having no cake. A *social system* such as the ones they propose is a *normative system* where policies and procedures are created in order to ensure equilibrium. In the cake example, the social system may add a policy that will make it possible for any agent that wants to eat a piece of the cake to do it.

[11]In fact the populations in Moses and Tennenholz's model are agents which are completely equal. But as Dawkins states in [51], in most societies there are more than one kind of population, each one with a different strategy. The success of those strategies depends not only in the number of different populations but also in the number of members of each population.

- In this model either the system can see the agents' goals in some way to ensure that all of them are inside the set of socially acceptable goals, or the model is assuming that the internal structure of the agents is such that it never will choose goals outside the G_{soc} set.

A Sociological Approach to Social Order

There is a school in Sociology known as *structuralism* that studies societies by creating methodic structural models to explain the different aspects of societies and social order. One of the works of this school that has influenced researchers in Computer Science is the one by Balzer [11]. In this work he presents a quite precise model of what he calls *social institutions* which comprises four dimensions:

- the *macro-level*, composed by *groups*, *action types* and *characteristic functions* to assign a collection of action types to a group;

- the *micro-level*, describing the individuals and the actions;

- the *intellectual representations* that individuals make from the macro- and micro-level structures, either in their minds or in the language they use for communication;

- the *social practices*, which are sets of actions that are performed by some groups and are imitated by other groups.

The *macro-level* of the social institution is called *core*. The core is defined as a structure

$$C = \langle \Gamma, \Theta, \chi, \lhd \rangle$$

where:

- Γ is a set of *groups*. Each group is characterized in terms of their behaviour and their status.[12]

- Θ is a set of *action types*. Action types are classes of actions which are similar in certain respects.[13]

- χ is the *characteristic function* that assigns *action types* to *groups*.

- $\lhd$ is a binary *status relation* among the groups, in order to indicate that a given group has *higher status* than another. The concept of status is closely linked to the notion of power defined in the micro-level.

The *micro-level* of the social institution is comprised by *micro-bases*, each of them defined as a structure

$$MB = \langle J, A, \text{PERFORM}, \text{INTEND}, \text{POWER} \rangle$$

where:

---

[12]The concept of *groups* in Balzer's model is the most similar concept (but not equivalent) to the one of *roles* in other models. A *group* is a collection of individuals that can perform the same action types.

[13]Balzer does not explain how *actions* are grouped together to form *action types*.

- J is a set of individuals,

- A is a set of concrete actions (that Balzer calls *action-tokens*) that individuals perform.

- PERFORM, INTEND and POWER are relations between individuals and actions:

    - i PERFORMS a: means that individual i performs an action a;

    - INTENDS(i,j,b): means that individual i intends that j should do b;

    - POWER(i,a,j,b): means that individual i by doing a exerts power over individual j so that j does b. Individual i is called the *superordinated agent* while individual j is called the *subordinate agent*.

The *intellectual representations* of the social institution are those mental frames[14] that get internalized by individuals and, if the institution lasts more than one generation, are then transferred to other individuals through the language. Such mental frames cover items such as *language*, *beliefs*, *dispositions*, and representations of the *core* components (such as *groups*, *action types* and the *characteristic functions*).

Finally, the *social practices* are set of actions that emerge, spread and are then accepted and imitated by other individuals of the social institution. They are modelled by predicates that state which individuals are the source of the social practice (SOURCE(γ)) and which individuals imitate them (COPY(γ)).

With all these elements defined, Balzer then refines the POWER relation as follows:

i by doing a exerts POWER over j to do b if and only if the four following requirements are satisfied:

a) actions a and b are actually PERFORMED by i and j,

b) i INTENDS that j should do b and j does not intend to do b,

c) the individuals believe that action a partially causes b,

d) i and j are members of groups γ, γ' such that actions a and b are admitted for i and j as members of groups γ and γ', and such that the admissibility of a and b for these groups is represented in i's and j's superstructure.

The POWER relation creates a network of ties between the individuals exerting power one to the other. It is interesting to note that, in the POWER definition, we can only say that an individual i exerts power over another individual j if j did not intended to do the induced action by itself. Another interesting idea is the one of *admitted actions* for a given group that restricts the actions that an individual may perform. Both the POWER relation and the *admitted actions* are the ways Balzer proposes to model the restrictive and prescriptive aspects of norms. However, Balzer does not explain how the POWER relations are created: the definition of POWER relation is presented in terms of an external

---

[14]Balzer refers to these mental frames as *superstructures*.

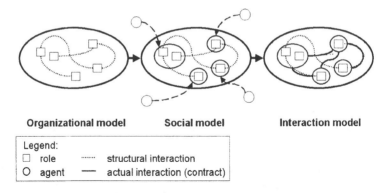

Figure 3.1: The three models in OperA (source: [58]).

observer, that is, it only says when an observed action of an individual can be defined as exterting power over another individual, but not who empowers such individual to do so.

The model is further extended by Balzer and Tuomela in [12] in order to include concepts such as *obligations* and *rights*. In this extension a *norm* is seen as a *task right system* (that is, a system that defines the rights of an individual to perform actions). In order to do so, individuals can be assigned to *positions*, where each position defines a set of obligations and rights attached with that position. The difference between a *position* and a *group* is that groups only define the set of action types that individuals can perform, while a *position* empowers an individual with rights and also imposes some obligations.[15] Although this extension allows the definition of norms, it cannot be used in practice to model open systems, as it does not define:

- how norms emerge or who can create them: the norms in the model do not require to be officially stated;

- how norms are enforced: there is an assumption that all the agents have a mutual belief that other agents will act according to the norms.

An Organizational Approach to Social Order

There is a quite extensive research in organizational science and economics focused on how people solve coordination problems by creating (more or less efficient) organizational structures. Within this view, and organization is seen as a set of individuals which group together in order to achieve one or several common goals.

Some of the research in human organizations has been successfully applied to model social order in software agent societies. A good example that integrates previous works is the proposed by V. Dignum *et al.* [60][59][58]. They present the OperA framework, which is composed by three interrelated models:

---

[15]Therefore, a *position* in Balzer and Tuomela's model is equivalent to the concept of *roles* in other models.

- the *organizational model*, which describes the structure of the organization in terms of roles and goal distribution among roles. The model assumes that the goals of the whole society are the ones that determine the agent roles needed, the relations between roles and their expected interactions (for instance, a *buyer* is not expected to interact directly with a *seller* in an auction scenario, but with an *auctioneer*).

- the *social model*, which describes the enactment of roles by the agents. The enactment is regulated by *social contracts*, that is, contracts where the agent explicitly accepts to enact a given role or roles, committing itself to the fulfillment of the goals assigned to the role(s).

- the *interaction model*, which describes the interactions between the agents that populate the society in terms of *interaction contracts*, that is, the explicit commits agents do with other agents to perform a given action or to achieve certain state. These commitments are acknowledged by the organization, which may act as a third party in order to grant the fulfillment of the interaction contract or impose some kind of sanction to an agent that does not fulfill its commitments.

In order to give some guidance to the possible interactions that may occur inside the organization, the framework defines *interaction structures*. These structures define patters of interaction in a way that the social goals can be achieved. An interaction structure is not a state-by-state definition of the interaction (i.e., a *protocol*), but a lightweight definition of some steps (that are called *landmarks*) to be followed. For instance, an interaction structure for an electronic trading house may be defined as follows:

- 1. Agents entering in an electronic market should *register* first.

- 2*a*. Then, if they enact the seller role, they should register the goods they are willing to sell.

- 2*b*. If incoming agents are enacting the buyer role, they have to register their requests.

- 3 Then both sellers and buyers may negotiate.

- 4 Finally, they should settle the trade.

The difference between an interaction structure (such as the list above) and a protocol is that the latter specifies all the steps, one by one, of the process, while an interaction protocol only fixes some landmarks, giving autonomy to the agents to decide how they will achieve each of the landmarks. [16]

In summary, in the OperA framework, the behaviour of agents is regulated not only by the *social contracts*, which are related with the (quite static) definition of the expected interaction among roles done in the *organizational model*, but also by the *interaction*

---

[16]The difference between interaction structures and protocols is similar to the difference between treasure hunt contests and standard races. In most of treasure hunts, participants have to find their path to a final treasure by hints that are given at some checkpoints: in each point the participant receives a hint to find the next checkpoint. This is opposed to standard races where participants have to follow a fixed course from the start line to the end line.

contracts that emerge from the actual interactions of the agents, interactions that have lightweight restrictions in the form of the *landmarks* defined in the interaction structure. However, the framework does not explain how to model the norms that the organization's context may impose to the organization, and how they should be included inside the different levels depending on the norms' level of abstraction.[17].

Logic Approaches to Social Order

There are some formalisms that have studied different kinds of *normative influence* between agents or roles by extending existing formalisms. As we saw in Section 3.2.1, Santos, Jones and Carmo's modal action operators E,G,H allowed to express a kind of *normative influence* between agents. Governatori *et al.* [88][81] extend this idea with the concept of *channels of deontic influence* in order to express that a given agent either is directly involved in the fulfillment of a norm, or there is a chain of agents linked by normative influences such that one of the agents will fulfill the norm. In order to model these chains, they propose to add a new action operator (EI). EI_iA means that agent i *attempts to make it the case that A*, by creating a channel of deontic influence terminating with A. Formally, this operator is defined as follows:

$$EI_iA \equiv G_iOG_jA \vee \bigvee_{\substack{j \in Ag_{\leq} \\}}^{j \leq i} G_iOEI_jA$$

where $j \leq i$ represents that agent i has deontic influence over j, and $Ag_{\leq}$ is the relation between agents that defines the aforementioned channels of deontic influence. The formula above is a recursive definition that expresses that either an agent i ensures that agent j ensures A or, through the chain of deontic influence defined by $Ag_{\leq}$, there is an agent that ensures that an agent will ensure A.

It is important to note that, although Governatori *et al.*'s formalism is quite expressive to model the effect of deontic influences in the fulfillment of norms, both the deontic influences and the actions resulting from such influences are directly assigned to agents instead of roles. Therefore, it is hard to use such formalism to model complex organizations defined in terms of roles. In order to solve such drawback, Carmo and Pacheco [32][33][160] have extended this action logics in order to introduce *roles* in the formula:

$$E_{i:r}A$$

expressing that agent i is enacting the role r when it brings about that A.[18] This extension allows a proper formalization of quite stable groups of agents[19] in terms of the roles they enact, the defined deontic influences and the interactions among roles.

In human societies the norms, the normative influences between actors and the actions to be performed are defined in the context of an institution. In order to introduce

---

[17]We propose a solution to this problem in Chapter 5

[18]The full axiomatization of Carmo and Pacheco's approach, linking roles with the O, P and F deontic operators, can be found in [160].

[19]Carmo and Pacheco refer to these groups of agents as *Organized Collective Agencies*.

this idea in a formalism, all the action logics we have mentioned use the *counts-as* operator $\Rightarrow_s$ (or slight variations of it). This operator, defined by Jones and Sergot in [107], is based in Searle's concept of *counts-as* in the context of a normative system. When applied to action description, a conditional

$$A \Rightarrow_s B$$

says that action A counts as (has the same effect or generates) action B. Several examples can be found in law. In the Spanish Law about organ donation [180], if a donor carries with herself a will to donate her organs, this action counts as a consent of that donor. This can be expressed through the $\Rightarrow_s$ operator as follows:

$$carry(donor, will) \Rightarrow_{SpLaw} Consent(donor)$$

where $SpLaw$ refers to the Spanish Law context where this connection holds. The connection can be used also to connect more complex formula such as:

$$E_x A \Rightarrow_s E_y F$$

stating that, within the institution S, if there is a situation where agent x *sees to it* that A, then it *counts as* the situation where agent y *sees to it* that F.

The *counts-as* operator can be used to define the effect of a power relation between two agents. The most significant approach is the one by Governatori *et al.* in [88], where they define the concept of *declarative power* of an agent as follows:

$$DeclPow_i A =_{df} proc_i A \Rightarrow_s E_i A$$

that is, an agent i has the *declarative power* of A if proclaiming ($proc$) that A counts as agent i bringing it about that A. The $\Rightarrow_s$ is a modification of Jones and Sergot's $\Rightarrow_s$ operator where some of the axioms are modified. The concept of *declarative power* can be also applied to link norms and power. For instance, we can express that agent i has the power over j to ascribe the obligation to achieve A in terms of declarative power as follows:

$$DeclPow_i OG_j A = proc_i OG_j A \Rightarrow_s E_i OG_j A$$

It is important to note that the declarative power of an agent is dependent on the institution s. Is the institution s the one which empowers the agent, as its power only holds within the context of s, by the definition of the $\Rightarrow_s$ operator. To finalise the framework, they formally connect the declarative power of an agent with the definition of a power relation ($\prec$) between agents (similar to the POWER relation defined by Balzer) as follows:

$$DeclPow_i OG_j A \Rightarrow_s i \prec j$$

In summary, there are some theoretical contributions that try to formalize the concept of power with the concept of norms. However, there are two main reasons that make it difficult to implement these approaches into agent architectures:

- different approaches choose different subsets of axioms to define the E,G,H and $\Rightarrow_s$ operators, so the contributions of each proposal are hard to be merged in order to, for instance, have a full axiomatization of Governatori *et al.*'s concept of *declarative power* formally linked with the concept of role enactment by Carmo and Pacheco.

- although these formalisms try to model actions by means of modal action logics, those actions are described in a very abstract way: agents are said to ensure or to bring about an action or a state of affairs, but not how that can be accomplished. For instance, a formula such as OH_jA states that j has an obligation to attempt A, but it does not say how the institution can prove that such agent attempted it or not. [20]

3.3.2 Discussion about Social-Centric Views

As we have seen, there are different approaches to analyze and model norms from the agents' society perspective. In these approaches the aim is to model a given society or organization by defining some kind of (social) structure that establishes the (accepted) relations among agents or roles. In some approaches such structure is constructed by means of defining *roles*, which define the restrictions to be followed by the agents that enact such roles. Some other approaches create the social structure by an accurate definition of the obligations to be fulfilled by a given agent and the relations of deontic influence between agents. However, in both cases the resulting social structure has a quite static nature, as it tries to identify and formally specify all the needed relations in a given setup, giving almost no option to create new ones dynamically (an exception to this statement is the OperA framework). Therefore, these approaches are not well-suited for open systems with heterogeneous agents.

Castelfranchi already identified this problem in [35]. He defines the aforementioned approaches as a *top-down* social control process, from the (static) specification of the social structure to the agents. He also proposes to include a *bottom-up* perspective, in order to model dynamic, emerging conventions and norms in a group of agents or the informal stablisment of rights, permissions and duties that appear in agent interaction. However, most of the work presented on this respect by Castelfranchi, Conte and the group at IP-CNR is mostly theoretical, as they have not presented yet an agent architecture integrating the top-down (formal) and bottom-up (informal) perspectives which can be used in practice.

3.4 Building Electronic Institutions. Multi-Agent Methodologies and Frameworks

The research on electronic institutions (*e*-institutions) focuses its efforts in finding formal approaches to specify multi-agent systems and their normative framework which are both

---

[20]We present our solution to this problem in Section 4.4, Section 5.4 and Section 6.4.3.

- *Expressive*: the specifications should use a formalism expressive enough to properly model *institutions, norms, roles* and other related concepts such as *contracts, commitments, obligations, rights, permissions, responsibility* and *authority*.

- *Useful for agents*: the norms that rule an institution have to be computable by an agent, so they can be included in practice in the agent's reasoning cycle, guiding completely or partially its behaviour.

As we have seen in previous sections, there is quite a lot of theoretical research on theoretical approaches and methodologies to model and design agent-based social systems. However, there are few frameworks and methodologies for *e*-institutions that have the properties mentioned above and that have been successfully used in practice to design and build a final multi-agent system.

3.4.1 Agent Methodologies without Normative Aspects

As our aim is to properly define normative frameworks for agents, we should discard methodologies that, although they are quite extended, are not well-suited to model the normative aspects of an agent society.

GAIA [227] is one of the first agent-oriented software engineering methodologies that explicitly takes into account some social concepts. GAIA aims at providing a coherent conceptual framework for the analysis and design of multi-agent systems. The analysis phase results in 1) the agent model which specifies system roles and their characteristics in terms of permissions (the right to exploit a resource) and responsibilities (functionalities); and, 2) the interaction model, which captures the dependencies and relations between roles by means of protocol definitions. GAIA is only concerned with the society level, does not capture internal aspects of agent design, and the implementation level is explicitly and purposefully ignored. Societies are only considered from the perspective of the individual participants, and therefore GAIA does not deal with communication or other collective issues. Normative aspects are reduced to static permissions (a sort of constraints or rules) and behavior is fixed in protocols. Furthermore, Gaia is not suited to model open domains, and cannot easily deal with self-interested agents, as it does not distinguish between organizational and individual aspects, and does not provide capabilities for agent interpretation of society objectives, norms or plans.

SODA [153], is actually an extension to GAIA that enables open societies to be designed around suitably-designed coordination media, and social rules to be designed and enforced in terms of coordination rules. As GAIA, SODA distinguishes between an analysis and a design phase. The analysis phase results in three different models: the *role model*, describe the goals, or tasks of roles and groups; the *resource model* that described the environment in terms of available resources; and the *interaction model*, where interaction protocols describe the information required and provided by roles and resources, and the rules governing interaction. During the design phase, roles are mapped to agent classes (the *agent model*), groups are mapped into societies designed around coordination abstractions (the *society model*), and resources are mapped into infrastructure classes (the

environment model). SODA provides some notion of context, or environment, of the society, albeit not explicit. However, even though SODA distinguishes between agent and
collective spaces, it sees roles as the representation of the observable behavior of agents,
and therefore cannot represent the difference between a) the organizational perspective
on the activity and aims of individuals, and b) the agent perspective on its own activity and aims. Role enactment is fixed in SODA as the agent model that maps roles to
agent classes without any possibility to accommodate agent preferences or characteristics
(agent classes are pure specifications of the role characteristics). There are no normative
aspects in SODA further than the notion of permission to access infrastructure services.
Communication primitives are limited to interaction protocols, and SODA provides no
explicit representation for the domain ontology. Furthermore, SODA also does not have
a clear and formal semantics.

MASSIVE [120] [121] has chosen an organizational perspective, and it has been
successfully used to model multi-agent systems which are then implemented in INTER-
RAP [71]. The methodology is inspired, in some extent, in object-oriented analysis and
design, in order to systematically obtain requirements that are then translated in agent
specifications which can be then implemented. It also defines a role hierarchy to model
the social structure of the agents in thesyste. However, MASSIVE is not adapted to institution modelling, as it can only represent some basic restrictions, which are attached to
the role definition.

TROPOS methodology [37] spans the overall development process. It distinguishes
between an early and a late requirements phase, and between architectural design and
detailed design . TROPOS starts with an analysis of the organizational setting of the
application, in the early requirements phase. The strategic dependency model describes
an 'agreement' between two actors: the depender and the dependee. The strategic ratio-
nale model determines through a means-ends analysis how an actor's goals (including
softgoals) can actually be fulfilled through the contributions of other agents. These two
models serve as input for the late requirements phase where a list of functional and non-
functional requirements is specified. The architectural design defines the structure of a
system in terms of subsystems that are interconnected through data, control and other
dependencies. The detailed design defines the behavior of each component. The models
are implemented using Jack Intelligent Agents [99], which is an agent-oriented extension
of Java. TROPOS is a fairly complete methodology that considers all steps in system development, and it treats both inter-agent and intra-agent perspectives. However, the late
requirements phase does not provide explicit concepts to capture norms. At the Implementation Level, TROPOS provides a detailed implementation of organizational models
into JACK agents. The main two shortcomings of TROPOS are that a) it is not formal (although there is some ongoing work on providing a formal semantics for TROPOS), and b)
it is too organizational-centered in the sense that is does not consider that agents can have
their own goals and plans, and not just those coming from the organization. Furthermore,
Tropos has no concept representing the normative aspects of an organization.

3.4.2 Agent Methodologies with Normative Aspects

Surprisingly, now-a-days there are very few methodologies and frameworks that apply expressive explicit representations of norms on the design and implementation of agents. And from those, we should discard the approaches that have been designed specifically for the e-commerce and e-business fields, such as SMACE [125] or the Contractual Agent Societies (CAS) [53]. In both cases the normative perspective is modelled by the use of *contracts*, which enforce agents to comply with the agreements they have done. But these contracts only model agreements between parties, and do not cover those permissions, prohibitions or obligations that an *e*-organization may need to define in order to ensure an appropriate behaviour of the society of agents as a whole.

A sound proposal is the SMART agent architecture presented by d'Inverno and Luck [61]. SMART is based in an agent specification framework developed in the Z specification language [202]. The framework defines concepts such as *objects*, *agents* (which are objects with goals) and *autonomous agents* (which are agents with motivations). The framework has been recently extended by López y López, Luck and d'Inverno [130][129] in order to introduce, as part of the framework, representations of norms. By studying the characteristics of normative MAS the authors aim to find the basis for a framework to represent different kinds of agent societies based on norms. An interesting contribution is that, in this framework, norms are not modelled as static constraints but as objects that can have several states (such as *issued*, *active*, *modified*, *fulfilled* or *violated*), and are related not only with the agents that should fulfill it or enforce it, but also with agents such as the one that issued the norm, the one(s) responsible of its enforcement (the *defenders*, the one that modified it or the one(s) that may be affected by a non-compliance of the norm. They also have analysed in [128] the different power relations that may arise in an agent society, not only the where the social structures define norms that entitle agents to direct the behaviour of others (*institutional power*) but also the power of an agent given by its capabilities to satisfy goals and the power of other agents to benefit or to hinder those goals (*personal power*). Moreover, the authors have also presented an architecture for autonomous social and normative agents in order to reason about norms. However, as far as we know, no implementation of the architecture applying it to a real problem has been reported in literature, and there are no tools to support the development of a normative multi-agent system following their framework.

The most contrasted proposal that we are aware of is ISLANDER [149, 188, 68]. This is the formalism we used to design the *CARREL* institution in Chapter 2. It views an agent-based *e*-institution as a type of *dialogical system* where all the interactions inside the institution are a composition of multiple dialogic activities (message exchanges). These interactions (that Noriega calls *illocutions* [149]) are structured through agent group meetings called *scenes* that follow well-defined protocols. This division of all the possible interaction among agents in scenes allows a modular design of the system, following the idea of other software modular design methodologies such as the Modular Programming or Object Oriented Programming. A second key element of the ISLANDER formalism is the notion of agent *role*. Each agent can be associated to one or more roles, and these roles define the scenes the agent can enter and the protocols it should follow. Finally, this formalism defines an easy to understand graphical notation that allows not only to obtain visual representations of scenes and protocols but also are very helpful while developing the final system, as these representations can be seen as blueprints. ISLANDER has been mainly used in *e*-commerce scenarios, and was used to model and implement an electronic Auction house (the *Fishmarket*). The resulting agent-mediated auction house was extensively used, for instance, to compare different strategies of the trading agents [49][80][133]. Furthermore, the eInstitution platform [66] and the ISLANDER API [103] enable the animation of models and the participation of external agents. The activity of these agents is, however, constrained by *governors* that regulate agent actions, to the precise enactment of the roles specified in the institution model. Furthermore, ISLANDER has a formal semantics [149, 188].

However, as we saw in Section 2.6, there are some drawbacks that limit the application of the ISLANDER formalism to complex domains. Most of them are caused because it models *e*-institutions in a low level of abstraction –which hardly can express complex, abstract norms that are present in human laws and regulations– by means of constraints for scene transition and enactment –the only allowed interactions are those explicitly represented by arcs in scenes–. Agents are not allowed to break these restrictions in any way, as Governors check all the messages that external agents utter and filter those that do not follow the specified protocol.

3.4.3   Electronic Institutions *vs.* Electronic Organizations

At this point, we want to remark the distinction that North makes between *institutions* and *organizations*. While institutions are abstract entities that define sets of constraints, organizations are instances of such abstract entities. The parties are members of an organization (not members of an institution) that should follow the *institutional framework* defined inside the organization. However, most of the work on *e*-institutions [72, 68] and even our previous work [48, 214, 56] use the term *e*-institution to refer to the agent architecture. From now on, we will keep as possible the accuracy in the use of terms, so we will refer to *Electronic Organizations* (*e*-organizations) as computational models that try to emulate a given organization and which follow an *institutional framework*. In Section 5.11 we will address this topic.

3.5   Summary

In this chapter we have seen how *norms* are viewed differently in different branches of Science. Then we reviewed some of the approaches to formalize norms and then how to introduce them in agents, either from an agent- or social-centric point of view. Finally we have shown some approaches to build *e*-institutions.

One of the conclusions from this review is that the foundations of the research in institution formalization come from several branches of science such as Sociology (and Socioeconomics) and Legal Theory, with several studies about the modelling of normative frameworks and the impact of such frameworks –that North calls *institutions* [151]– in

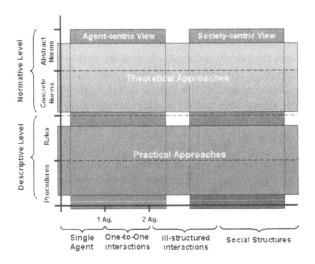

Figure 3.2: Two axis to classify the different approaches.

societies and organizations. However, the *electronic institution* concept is quite new, and formalisms to fully describe all the aspects involved are still to be developed. As Balzer and Tuomela point out in [12]:

> *"A comprehensive theory of institutions is still missing which makes explicit the overall macro structure, the norms, and the systems of actions as well as the interplay between these components".*

The lack of a unifying theory that covers all the aspects needed to fully model societies and organizations following a given normative framework is due, in part, to the aforementioned variety of foundations of the research made. Although different proposals cover the normative perspective, the static social structure and the (emerging) dynamic relations among individuals, they are hard to combine in a single model because of the assumptions coming from different areas (Social Theory, Organizational Theory, Decision Theory) that are part of these proposals.

In order to summarize here the different approaches that can be found in literature, a two-axis view is proposed in figure 3.2:

- The horizontal axis describes the *social dimension* of the approaches, ranging from the simplest, *single agent* ones (where interaction is not studied), the *one-to-one interactions* (interactions are statically defined and modelled), the *ill-structured interactions* (interactions can be emerge dynamically, but follow no organized structure) to the *social structures* (interactions are socially structured around some global objectives).

- The vertical axis describes the *normative dimension* of the approaches. It uses the same levels that are defined in Chapter5 for the HARMON*IA* framework. Therefore, we make a distinction between the *normative level* (the one that states what is acceptable or unacceptable by means of *obligations, permissions* and *rights*) and the *practical, descriptive level* defining the *praxis* (that is, how agents should behave in order to meet the norms). In the former, *values, abstract* and *concrete norms* are used to model a normative system in a highly expressive way, while in the latter, *rules* and *procedures* describe how the norms are to be interpreted and used (operationally) by the agents.

By means of these two axis, we can graphically locate the different families of approaches that we have presented in this chapter. The horizontal axis allow to distinguish between *agent-centric* and *social-centric* views, while the vertical axis allows to distinguish between *theoretical* and *practical* approaches. Figure 3.3 depicts the approximate areas that some of the representative methodologies and approaches cover.

- On the top-left corner we can find the most formal approaches such as *Standard Deontic Logic* (*O,P* and *F* operators) and its extensions, the *Logic of Indirect Actions* (*E, G* and *H* operators) by Santos, Jones and Carmo [190][191] and the concept of *Deliberative Normative Agents* [36]. In all these cases the approaches are focused on the expressiveness of the logic and/or the consequences in the reasoning cycle of a single agent, covering at most the obligations that one agent may have towards another.

- On the top-right corner we can find approaches such as the extension of the *Logic of Indirect Actions* with *roles* ($E_{x:a}A$) by Carmo and Pacheco [33][160] and frameworks such as OperA [59][58]. In this case focus is given to model the organizational aspects of the society, mainly by the definition of roles, the relations among them and the norms that apply to these roles. However, these approaches are still theoretical and have no implementation counterpart.

- On the bottom left corner we can find formal approaches such as GAIA [227], where agent behaviour is not constrained by normative constructs but by simpler, descriptive rules and constraints.

- On the bottom right corner we can find methodologies such as TROPOS [37] and ISLANDER [188][68], where the normative expressiveness of the formalism is reduced in order to be able to create implementable solutions (normative aspects are reduced to the definition of some rules and constraints). These methodologies both cover, in some extent, the agent and the society view.

It is interesting to note that the different approaches are either in the normative or in the descriptive level. This is because each of these approaches try to model all the normative aspects with a single formal language. In the case of the more theoretical approaches, the formalisms and theories are usually based in *possible worlds* semantics. As we mentioned in section 3.2.3, Wooldridge and Ciancarini state in [224] that with this

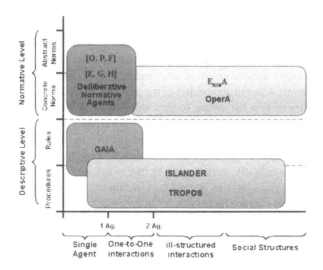

Figure 3.3: Some of the reviewed approaches, classified within the two axis.

kind of semantics there is no precise connection between the abstract accessibility relations that are used to characterize an agent's state and any concrete computational model. Therefore this makes it difficult to go from a formal specification to an implementation in a computational system. In the case of the more practical approaches, a formalism with operational semantics is chosen, expressing the norms as behaviour constraints.[21]

Although this variety of approaches, most of the definitions of institutions are based in the (more or less formal) definition of their *norms* and their effect in the actions performed by the members (the entities that belong to the institution and which follow its norms). As mentioned in Section 1.3.3, those norms, instead of being an obstacle, are helpful for the agents as they:

1. decrease uncertainty about other agents' behavior inside the institution,

2. decrease misunderstanding, as there is a common corpus of norms governing agent interaction,

3. permit agents to foresee the outcome of a certain interaction among participants

4. simplify the decision-making process inside each agent, by decreasing the number of possible actions.

---

[21]As we will see in Section 5.1.1, in order to avoid this dichotomy between normative and descriptive formalisms, we propose to use more than one formalism to describe the same system, choosing proper formalisms for the normative level and the descriptive level, and defining a formal connection between them.

All the points are very useful from the Multi-Agent Systems' perspective, as they simplify the coordination problem, and allow the agents to perform more accurately despite its bounded rationality.

As we have mentioned, one of the main issues is to decide how norms are to be used within the context of multi-agent societies and institutions and how the agents should cope with these norms. Current approaches of the problem are opposed. Some approaches hard-code the restrictions in the agents, so they will always behave according to these restrictions, without knowing that these restrictions follow from specific norms. Therefore, with these approaches relatively simple and small agents can be used, as the overhead caused by reasoning about the possible consequences of their behavior is minimum. Another approach is to explicitly represent the norms of the institution. Therefore, in this case more complex agents have to be designed in order to be able to reason about the norms and the possible consequences of every action, adapting their behaviour to special circumstances (e.g. the ones when it pays off to violate a rule).

As we will see in Section 5.1.1, we propose to have a link among procedures and the related norms. This allows having quite simple agents that usually follow the procedures but, whenever they want, can obtain the related norms to reason about them and make better choices. We will call this kind of agents *Flexible Normative Agents*. In this scenario, the *e*-organization that will function according to some explicit norms (the institutional framework) do not force a huge overhead on the agents that operate within that institution. They only create the possibility for more intelligent behavior of the agents that are capable of coping with such an overhead.

Part II

HARMON*IA*: a New Approach to Model Electronic Institutions

Chapter 4

A Multi-Layered Normative Framework for Agents

In Chapter 3, different approaches to model agent societies have been presented, from those coming from Social Sciences or Legal Theory to those in Computer Science. We have also seen the different views of the term *norm* in different branches of science, and more important, the existing gap between theoretical and applied approaches to the use of norms in Multi-Agent systems.

Hence, some kind of unifying framework is needed. In this chapter and the following we introduce HARMON*IA*[1], a framework that defines a multi-level approach, from the most abstract level of the normative system to the final implementation of an *e*-organization. Our framework does not pretend to unify all the aspects of the modelling of agent societies –that would be too ambitious– but in the normative aspects, which have a prominent role in the definition of *e*-institutions.

This chapter focuses on the study of norms from the agent point of view. The relation among the norms and the agent's beliefs, desires and intentions will be described. We will first check the effect of the different levels of abstraction that norms present from the epistemic point of view, and then we will see how the norms modify the standard BDI process. By doing so we will identify some of the elements that compose our framework (*norms, rules, procedures, policies, roles, context*) and we will provide an accurate definition. The proposed terminology attempts to unify, as much as possible, the different contributions in the study and modelling of normative frameworks, and it aims to ease future advances in the field by providing a kind of standardization of the terminology, so results of different research teams can be merged.

The view of our proposal will be completed in Chapter 5, where we will study the norms from the institutional point of view. By the end of the next chapter all terms will

---

[1]The name comes from *Harmonia*, Greek Goddess of harmonious relations among parties, known by romans as *Concordia*. Daughter of *Aphrodite* (Goddess of beauty - *Venus*), surprisingly she was also daughter of *Ares* (war - *Mars*) and sister of *Eris* (strife - *Discordia*), *Phobos* (panic) and *Deimos* (fear).

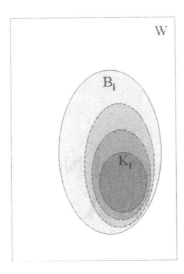

Figure 4.1: Beliefs and knowledge of agent a_i about the world W.

be defined and the complete our proposed framework presented. Then in Chapter 6 an extensive example will show how all these concepts are applied to a real scenario.

4.1 Intuitive Description of the Impact of Norms in Agents

In order to understand how the norms modify the reasoning process of an agent, let us first analyse the relation among the norms and the agent's knowledge about the world. To do so we will depict such relations by means of Venn diagrams representing the epistemic dimension.

The model presented in this section is an intuitive first version that tries to capture our intuitions about the impact of norms in the agents' knowledge. In Section 4.2 a more formal approach (based on the idea of *possible worlds*) is used to complete this model.

4.1.1 Knowledge and Beliefs

The kind of agents that interest us are those which base their decision making process in the knowledge they have of their environment.

Figure 4.1 depicts, from an epistemic point of view, the perception an agent i has of its environment:

- W represents the world, the environment where the agent is situated,

- B_i is the set of beliefs the agent has,

- K_i is the subset of the beliefs the agent knows for sure.

Intuitively, the set B_i corresponds to the $\Diamond$ or BEL operator in Epistemic logic, while the set K_i corresponds to the $\Box$ or K operator. However, in this first intuitive model we will not establish any logic to do the description, in order to avoid the restrictions some of them impose.

In fact, the set B_i has several certainty levels, from the lowest ones, closer to the border of the set, to the higher ones, where the agent is so certain that they are considered knowledge instead of beliefs. Figure 4.1 also represents this fact. Although in the figure there are only depicted four levels, such division is only made for display purposes, as the set B_i should be considered as a continuous certainty field[2].

During the life of an agent, the certainty of the facts contained inside B_i can increase (if there are new facts that support them) or decrease. As K_i represents the agent a_i's knowledge, changes in the border of the K_i set correspond to the learning process of the agent:

- The *growing* of the K_i set corresponds to the AI classical view of symbolic learning. As noted by *Martin's law*[3], only the facts the agent is almost certain can be learned, that is, the facts that are closer to the border of the set K_i.

- The *shrinking* of the K_i set corresponds to cases where (i) a change in the world decreases or changes completely the set of beliefs and Knowledge (in classical Epistemic Logic); or (ii) the rise of new facts may decrease the certainty of part of the knowledge of the agent (in some non-classical reasoning approaches such as non-monotonic, possibilistic or probabilistic reasoning)[4]

Even though the set K_i changes its size, it always is true[5] that $K_i \subseteq B_i$.

4.1.2 Goals

Goals are one of the central concepts in Agent Technologies. They represent the states or actions the agent aims at. Although there are several definitions of *Agent*, all of them agree on the need of defining goals.

Even though, from an operational perspective, *goals* are part of the decision making mechanism of the agent and therefore, *goals* are more related with actions and tasks than with the epistemic dimension, there is an intuitive relation among goals and the beliefs and knowledge. Figure 4.2 shows such relation: lets define the set G_i as the set of goals

---

[2]With this continuous interpretation of B_i, the frontiers of the sets shown in figure 4.1 are certainty isolines.

[3]The Martin's Law states: *"You cannot learn anything unless you almost know it"*. This means that the learning process usually should be done in small steps, otherwise for each step there is too much to figure out, and that gives more room for error and confusion.

[4]In fact, in some of those approaches there is no concept of Knowledge, but only certainty levels in Beliefs that can increase or decrease.

[5]The case $K_i = B_i$ is the case when the agent works with only two levels of certainty ($true, false$).

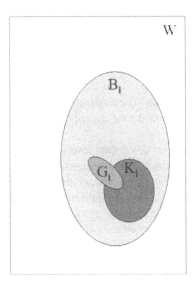

Figure 4.2: Goals of agent a_i.

of a given agent a_i that it believes that are possible, given its beliefs and knowledge about its capabilities and the current state of the world.

An example could be the following:

> *"Imagine that John wants to offer a collar to his wife as a present. John knows which one to choose (a beautiful and expensive one that both John and his wife saw some days ago). John knows the place and also that it closes at 18:00. Checking the cash he has in his wallet he also knows that he has not enough cash. He has an AMEX and a VISA card, and he has a very certain belief that such shop accepts credit cards, as he remembers the stickers on the door. With all this in consideration, John adds this task as one of the things to do during the day (a goal for the day), as he thinks that it is possible to buy the collar".*

Summarizing, in the example the agent a_i (*John*) has some beliefs and knowledge about facts that *ground* his decision to add `buying the collar` as a goal. This intuitive idea is represented in figure 4.2, where the set G_i overlaps the sets B_i and K_i, as the goals in G_i are based on facts that belong to the beliefs and knowledge of a_i. It is important to note here that this definition of goals is completely different from the one in BDI logics, where goals are expressed in terms of states[6]. So the diagram does not intend to mean that the set of goals is inside the set of beliefs ($G_i \subseteq B_i$), but that the goals are

---

[6]In BDI logic, the formula $G_i(\phi)$ means *"i has the goal to achieve a state where ϕ holds"*. As we will see in section 4.2, with this definition the intersection between K_i and G_i is senseless, as one agent should not have ϕ as goal if it is already true.

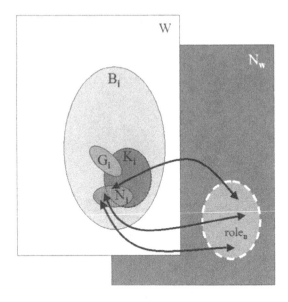

Figure 4.3: Knowledge and beliefs about the norms.

grounded by[7] the beliefs. A more formal definition of goals is presented in Section 4.2.

4.1.3 Beliefs and Knowledge about Norms and Roles

Apart from the beliefs and knowledge that the agent has about the world, in the case of *e*-institutions it is also very important to view the beliefs and knowledge about the norms that apply in that world.

Let us define N_w as the set of norms (the *institution*) related to the world W (the *organization*). Such set of norms comprises all the normative framework of all the kinds of agents that interact in the organization, agent a_i being only one of those agents.

As not all the norms apply to all the agents, agent a_i only needs to know about a subset of N_w. But a direct mapping from norms to agents is not acceptable, as norms are defined at the design phase of a system from the society point of view, and in such a step it is not a good methodological option to consider which are the particular agents that will follow a given set of norms.

In order to get some independence between the norms and the concrete agents, a definition of roles is mandatory. *Roles* represent the different entities or activities that are needed to fulfil the social purpose of a system [59]. From the agent's perspective this

---

[7]Even though we are defining in this section a first intuitive model, the *grounded by* relation can be translated into an `achievable(`$a_i, goal, state$`)` function expressing that the goal *goal* is achievable for agent a_i given the current state of the world *state* and the capabilities of that agent.

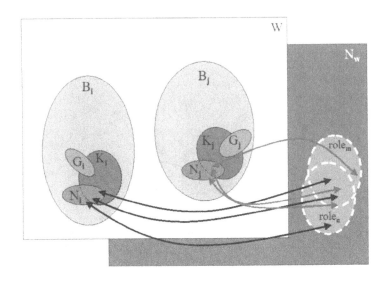

Figure 4.4: Shared interpretation of norms among agent a_i and agent a_j.

means that agents not only have intentions that guide their actions but also they should have the ability to enact one or several roles. From the social perspective the definition of roles is a mechanism to design a society in a way that the global goals are covered by the roles defined, without being concerned about the specific agent or agents that will play a specific role at runtime, as long as such role is performed. Depending on the way a system is described, roles are usually defined in terms of the goals to be fulfilled [163], the capabilities that the role should have, the protocols to be followed [188, 68] or, as in our case, the norms that should guide –not to impose– its behaviour.

With this view let us define $role_n \subseteq N_w$ as the set of norms that apply to a given role (role$_n$) that agent a_i is playing in a given moment (the set $role_n$ is represented in light grey in figure 4.3).

As the sets B_i and K_i define the perception agent a_i has of the world, the set N_i defines the perception agent a_i has of the norms it has to follow according to its role. As in the case of the set G_i, the relation of the set N_i with B_i and K_i is not an inclusion one but the following: the perception that agent a_i has of the norms may be grounded either by facts in the set B_i and/or in K_i.

N_i is, in fact, a model of the norms in the set $role_n$, and it results from a interpretation of the norms. The interpretation process is depicted in figure 4.3 by means of black arrows, as interpretation can be seen as a kind of mapping from the norms to the low-level actions and states the agent a_i perceives in the world W (as we will see in Section 4.2.1).

The set N_i has a prominent role in our model, as it represents the interpretation that agent a_i makes of the normative framework, and which guides its behaviour. Therefore, one of the main issues when designing e-organizations is to design mechanisms to ensure that all agents that perform a certain role role_k make an *equivalent* interpretation of the norms in $role_k$. The idea is that, even though the agents are the ones that decide how they enact the role, they should, at least, interpret the boundaries and constraints such norms impose. An example is depicted in figure 4.4, where there are two agents a_i and a_j situated in the same world W. Both agents have their own beliefs (B_i,B_j), knowledge (K_i,K_j) and goals (G_i,G_j). They have also their own model of the norms that apply to them (N_i,N_j) that comes from their interpretation (represented in the figure as arrows) of the sets of norms that are assigned to them $(role_n,role_m)$. As shown in the figure, the ideal situation is that the norms in the normative framework that are shared among agent a_i and a_j $(role_n \cap role_m)$ are similarly interpreted by both of them (this is represented in the figure by the intersection of the N_i and N_j sets)[8].

However, if interpretation of abstract norms is completely done by the agents, it is hard to ensure that two agents would perform the same or similar interpretation. Therefore, even though we tried to keep the model as simple as possible, there is a need to define some elements in such a model in order to reduce the number of possible interpretations of a set of norms to the acceptable ones from the point of view of the agents' society.

4.2 Conceptualization of Norms in Terms of Possible Worlds

To be able to properly refine the model presented in Section 4.1 and add some more elements, some kind of formalization is needed. As we will see in Section 5.2.2, we will choose modal logic (Deontic logic) in order to express norms because of the expressivity such approach gives to the normative framework. So, with such approach, it is natural to express an agent's mental states in terms of *possible worlds* that are the basis of *Kripke semantics*. This will also allow us to describe the agent's mental states by means of BDI logic and then, in Section 4.3, see how norms impact the BDI internal cycle of the agent.

As a first step we will redefine the set W as follows:

Definition 4.1. *W is the set of possible worlds w that define the environment of the agents.*

The redefinition of the environment W (depicted in figure 4.5) leads also to a redefinition of the sets B_i (*beliefs*, which corresponds in Epistemic Logic to the BEL operator), and K_i (*knowledge*, which corresponds in Epistemic Logic to the K operator). :

Definition 4.2. $B_i \subseteq W$ *corresponds to the set of worlds that the agent a_i believes with a given certainty level.*

---

[8]Only for simplicity purposes, in figure 4.4 there is no intersection among the sets $\{B_i, K_i, G_i\}$ and $\{B_j, K_j, G_j\}$. However, in some setups such as *cooperative problem solving systems* where agents have a set of common goals, such intersection is not only possible but also desirable, as agents should share goals and knowledge.

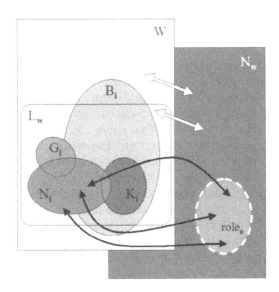

Figure 4.5: Redefinition of W, B_i, K_i, G_i and N_i in terms of possible worlds.

Definition 4.3. $K_i \subseteq B_i$ *is the subset of worlds w that the agent a_i knows with a high certainty level.*

The redefinition of the set K_i in terms of the modal operator K adds one constraint to the model: The set K_i will not be allowed to shrink, as axiom T states that $K\phi \longrightarrow \phi$.

Also a new definition of the set of goals is needed in terms of W:

Definition 4.4. *The set $G_i \subseteq W$ is the set of possible worlds the agent a_i has as objective to reach.*

It is important to notice here that we have abandoned the intuitive definition of goals presented in Section 4.1.2 towards the definition of goals in BDI logic ($G_{agent}(p)$). This redefinition brings at least two properties that have to be taken into account:

- As a result of definition 4.4, $G_i \cap K_i = \emptyset$, as it does not make any sense that an agent a_i has a goal $G_{a_i}(p)$ if a_i knows that proposition p is already true ($K_{a_i}(p)$). This change is reflected in figures 4.5 and 4.6.

- In order to define goal *achievability*, define the function $\texttt{achievable}(p, a_i, w_1)$, where $G_{a_i}(p)$ and $w_1 \in W$, to express that the agent a_i can achieve from the world w_1 the goal p according to its capabilities and its knowledge.

4.2.1 Norms as Legally Accessible Worlds

Finally, to complete the redefinition, we have to redefine the meaning of the set N_i in terms of possible worlds. To do so we will introduce the concept of *socially acceptable worlds* presented in the Artificial Social Systems' model by Moses and Tennenholtz [9]. In [144] they defined the set W_{soc} as the set of worlds that all agents can legally reach. Following this idea we define the set L_w as:

Definition 4.5. *The set $L_w \subseteq W$ is the set of legally accessible worlds for all agents that are situated in W.*

The Artificial Social Systems' model has no definition of roles. But in our model roles are a prominent concept (see Section 4.2.5). Now we are able to redefine in our model the set N_i in terms of possible worlds and roles as follows:

Definition 4.6. *The set $N_i \subseteq L_w$ is the set of possible worlds the agent a_i can legally reach according to the norms defined in the set $role_n \in N_w$.*

The relation among the world W and the norm set N_w can be formally defined as follows: let us also define a function MAP: $Norms \rightarrow W$ such that:

- $Norms$ is the language used to express norms in N_w,

- $L_w = \text{MAP}(N_w, W)$,

- $L_i = \text{MAP}(role_n, W)$

It is important to remark that, with this new definition of N_i, an agent should only have an option to fulfill its goals if they are socially acceptable (otherwise they will enter in a *violation situation* . In the Moses and Tennenholtz's model, they assume that the goals of the agents are visible in some way by the system. That assumption allows them to define the set G_{soc} as the set of *socially acceptable goals*. However, in our model we will not assume that the institutional framework of the *e*-organization will have access to the agents' goals, as the *e*-organization will see agents as black boxes. Therefore, the control of the behaviour of the agents will not be made in terms of *socially acceptable goals* but in terms of *socially acceptable worlds*. So, from the organization's point of view, an agent a_i should only be able to fulfill the goals in G_i that belong to the set $\{G_i \cap N_i\}$. But in real setups organizations cannot always enforce the fulfillment of all the norms in a strict way, so it may happen that agents achieve goals that lead to worlds that are not socially acceptable. In this cases a violation occurs. In real setups *violations* are really useful, as usually they define procedures to be followed in order to make the agent achieve a state that is legal. An example is a driver that becomes *illegal* because of exceeding the speed limit in a highway. There is no way for the Police Department to avoid people driving above speed limits. But when a policeman detects a driver violating a norm, then the driver becomes illegal until it follows the procedures to become *legal* again (e.g., paying a fine). Returning to the *e*-organizations environment, figures 4.3 and 4.4 are examples of agents that will always become *illegal* if they try to fulfill their goals,

---

[9]This model is reviewed in Section 3.3.1.

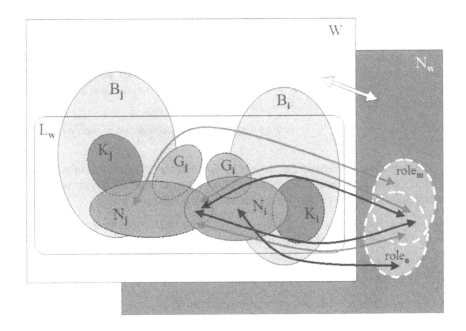

Figure 4.6: Beliefs, knowledge, goals and norms of two agents a_i and a_j in the same world, in terms of possible worlds.

as they will aim to reach worlds that are outside the set of worlds that they can legally reach according to their roles ($\{G_i \cap N_i\} = \emptyset$, $\{G_j \cap N_j\} = \emptyset$). Figures 4.5 and 4.6 are examples of agents that can fulfill at least some of its goals without becoming *illegal*, as some of the worlds it aims to reach belong to the set of worlds it can legally reach according to the role it plays (i.e., $\{G_i \cap N_i\} \neq \emptyset$).

We will talk again about enforcement of norms in Section 4.4.

4.2.2   Context

Now that we have re-defined our model we can include the missing elements that we referred to at the end of Section 4.1.3. One of those missing elements is *Context*.

In AI there are several systems which domain models suffer of what Guha calls *Homogeneity of the KB* [93]:

- the same vocabulary is used uniformly throughout the KB,

- KB contains a single theory of the world that must be kept consistent,

- the single theory approach should be kept generic and independent of particular problems, i.e., the representation should not be tailored specifically towards solving certain problems.

The problem that arises with the *single vocabulary/single theory* approach is that it tries to find a universal vocabulary, theory and representation to model and reason about any situation. The alternative, first proposed by McCarthy in [134], is to include context as formal objects.

There is not a single definition of context, as different branches of the AI use context for different purposes:

- In the area of Natural Language Processing (NLP) and Agent Communication, *contexts* define the meaning of the terms [14, 15][17][18][194][43] and the intention of a speech act [2].

- In the area of Cognitive Sciences concerning knowledge representation and reasoning, contexts define the validity of propositions and axioms [135]. A good example is the sentence $ist(c, \phi)$ that expresses that ϕ is true in context c [93, 26, 136]. Some examples of logics that handle context are *Local Model Semantics* (LMS) [85, 82], *Distributed First Order Logic* (DFOL) [83], *Propositional Logic of Context* (PLC) [25] *Contextual Intensional Logic* (CIL) [205] or *Context-Mediated Behavior* (CMB) [208].[10]

Penco [167] summarizes these views by distinguishing two views of context:

- context as an *objective, metaphysical state of affairs*: in this view a context is a set of features of the world needed to identify it (such as *time*, *place* or *speaker*),

- context as a *subjective, cognitive representation of the world*; in this view a context is a set of assumptions on the world expressed, for instance, by means of *axioms* or *rules*.

While the first view considers context from an ontological perspective, the second focuses on an Epistemic perspective.

The concept of *context* is important in our case as organizations (such as an Electronic Auction House) are examples of context of an agent. They not only define specific vocabularies but also norms to be applied within such organizations. This allows us to extend the distinction that North makes between *institution* and *organization* including context as follows:

Institutions are abstract patterns that are valid for all the environment W, while organizations are concrete instances of an institution defined inside a certain context C_a.

Therefore, most systems have moved to have an explicit representation of context. One of the most used approaches is the *box metaphor*, that is, considering context as a box:

[...] Each box has its own laws and draws a sort of boundary between what is in and what is out.[84]

---

[10]A good comparison among two of those logics (LMS and PLC) is presented by Bouquet and Serafini in [23].

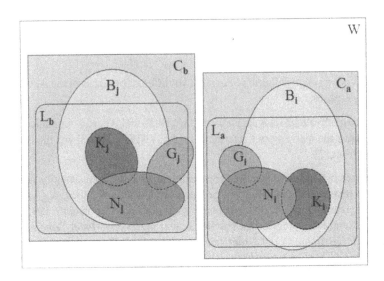

Figure 4.7: Two agents in the same world but in two different contexts.

With this idea, we may define context in our model as follows:

Definition 4.7. *A context C_a is a subset of worlds $w \in W$ where there is a shared vocabulary and a normative framework to be followed by a certain group of agents. A context may contain other contexts inside it.*

Figure 4.7 shows an example of contexts. For simplicity it only shows two contexts that are totally independent. Agent a_i is situated in a context C_a. The context C_a defines a normative framework, that in terms of possible worlds, it is the set of *physical accessible worlds* for all the agents inside C_a, agent a_i included. The sets B_i and K_i represent the knowledge agent a_i has about the context it is situated in. Based in that knowledge, it has also its set G_i of goals. The same applies for agent a_j, which is situated in the context C_b. We will talk about the sets N_i and N_j later in Section 4.2.3.

Definition 4.7 also refers to a vocabulary to be shared by agents in a given context. It means that each context is associated with a domain ontology that defines the meaning of the terms. A good example is to compare two contexts: the western society *versus* the muslim orthodoxy society. Apart from the well-known differences in faith and customs, we can find different interpretations of concepts that seem equally defined world-wide. One of such concepts is *family*. The western traditional view of *family* is of a man and a woman that form a couple (*monogamy*), and may have some descendants (their children). Sometimes the family is extended with one or more grandparents. The Muslim view of a *family* is similar to the western one, except that a family can be composed of a man with more than one women (*polygamy*), each one with its own descendants[11]. Another

---

[11]Of course these definitions of *family* are too simplistic and too archetypical. We will extend the definitions

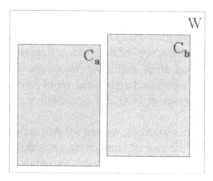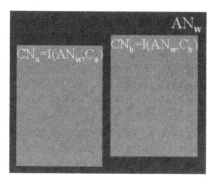

Figure 4.8: Parallelism among environment-contexts and abstract-concrete norms.

interesting example, taken from the science-fiction, is the three-membered alien families that Isaac Asimov describes in the award-winning novel *The gods themselves* [8]. In that alien race there are no two genders but three, and marriage is done not in pairs but in triples. The main difference in all contexts is the meaning of *marriage*, and as a result different family structures are found in each of these contexts.

In fact, in the western society context, we should extend the *family* concept in order to include other kinds of families (families only with a father or a mother, families where the parents are not married, families with both parents of the same gender, families where either the grandfathers, the godparents or the older descendants play the parents' role, etc.). So in this context we should define family as a group of people where some of them play the *caretaker* role, taking care of the others (the *cared*).

Another concept with a more or less world-wide definition is *monarchy*. In the last thirty centuries several ruling patterns have appeared where a pyramidal hierarchy is defined with a *ruler* on top that commands the people that are in the lower levels of the pyramid. Different contexts define different structures in the pyramid and different definitions of the *ruler*: from European *kings* to Asian *emperors* or Russian *czars*, from Egyptian *pharaohs* to Islamic *emirs*, from the *Great Mogul* in India to the native-American *chiefs*. We can also see different definitions of the *king*'s role in Europe through the ages: while in the middle ages the king was the commander of the army and used to be present in all battles, most of the modern European kingdoms (since the 17th century) used to send the army to the battle in the name of their king without the king's presence in the battlefield.

4.2.3 Abstract and Concrete Norms

In human societies a context may also have an associated normative framework. The normative framework defines a set of norms that apply in that context. This relation also

---

later in this section and the following.

happens in the opposite direction: Norms and Regulations usually define the *scope*, that is, the context where such norms and rules should be applied. Here a *box metaphor* approach can also be applied, as scope defines a boundary among the situations where a given Norm or Regulation applies and the ones that are "outside the scope" of the Norm or Regulation.

In figure 4.7 we have an example of the effect of the normative framework in the agents. While contexts C_a and C_b define, as we mentioned earlier, the set of *physical accessible worlds*, the sets (N_i or N_j) define the subset of *legally accessible worlds*, limiting the possible actions of agents a_i and a_j.

Returning to the western and Muslim societies example, we can see that both societies have defined norms. Such norms define in the case of families the cases that are acceptable or unacceptable in a given context. It may happen that a certain family pattern is accepted (*legal*) in one context and unaccepted (*illegal*) in the other. For instance, polygamy is legal in the Arab context, while it is illegal in the western one; and a single mother or a gay couple are *legal* in the western context and *illegal* in the Muslim orthodox one.

Despite the differences, in both contexts the norms are based on some common moral conventions, such as goodness or fairness. The *caretaker-cared* definition of family not only defines western families, as it is so abstract that Muslim families can also be included in it. So we can define an abstract level where we can define concepts, roles and norms (such as the one that says: *"caretakers must take care of the cared"*) that are universally accepted. Returning to our model, the abstract level of the normative framework is represented by the set AN_w:

Definition 4.8. *The set of Abstract Norms AN_w is the set of all possible norms that are valid for all the environment W. An institution is a subset AN_x of abstract norms in AN_w ($AN_x \subseteq AN_w$).*

The first part of the definition states that norms defined in the set AN_w are valid for all the worlds $w_i \in W$. The second part gives us a first definition of *Institutions* in our model, and place them at the abstract level. As explained in Section 3.4.3, institutions are abstract patterns that are instantiated by organizations. For instance, we can identify abstract concepts as *family* or *monarchy* as institutions, while my family or the Spanish monarchy are instances of those institutions (that is, the interpretation of the abstract concepts in some concrete contexts).

While *Abstract Norms* are defined in an abstract level, we find norms with lower levels of abstraction defined in a particular context (particular scope). Those norms (that we will call *Concrete Norms*) extend the norms in the abstract level and also define more precisely terms and roles defined in the abstract level:

Definition 4.9. *A set of Concrete Norms CN_a is the interpretation of the set of Abstract Norms AN_w for a certain context C_a. This can be expressed with the formula $CN_a = \mathrm{I}(AN_w, C_a)$.*

It is important to notice here that the interpretation function I should not be confused with the INTEND operator in BDI logic. Let us define $ANorms$ as the language to express norms in AN_w, while $CNorms$ denotes the language used to express norms

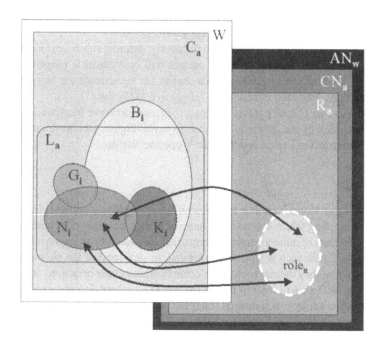

Figure 4.9: Relation of agent a_i with its context, abstract norms, concrete norms, rules and roles.

in CN_a. Function I: $ANorms \rightarrow CNorms$ is a mapping from the norms in AN_w to the norms in CN_w. [12]

A good example of how context defines an interpretation of norms is *monarchy*. At an abstract level we may define (i) terms such as *command, command chain, territory, population* or *hierarchy level*; (ii) roles such as *commander, ruler, servant, subject*; (iii) norms such as *"the commander commands its immediate servants"*, *"the ruler is the higher commander and is placed at the higher hierarchy level"*, *"all the population in the territory under the ruler's command are subjects"* or *"the command chain is a chain of command relations, with the ruler on top and a subject at the bottom"*.

When we choose a concrete context (e.g., a European kingdom in the middle ages) we (i) add more specific terms such as *kingdom* (for *territory*), (ii) define more concretely the roles, even introducing more categories such as *king, queen* (for *ruler*), *lord, count, duke, duchess* (for *commander*) or *vassals* (for *subject*), (iii) specify the norms in the abstract level in terms of the roles and terms defined in the concrete context, and add those norms that apply to European kingdoms which are not expressed in the abstract

---

[12]We will talk again about this mapping in Section 5.3, where we propose to use a variation of the *counts-as* operator $\Rightarrow_s$ to model the interpretation function.

level (for instance, *"All vassals who live in the kingdom enjoy the protection of their kings and owe allegiance to the king and the queen"*).

This refinement process can be done iteratively, defining subcontext inside the contexts until the context where a group of entities will be situated is properly identified. For instance, we can define a *duchy* or *dukedom* as a context inside the *kingdom* context, adding new terms, roles and norms (e.g., *"allegiance to the duke"*).

As we have seen, each context defines a concrete normative framework with a concrete interpretation of the norms in the abstract level. Figure 4.8 shows the parallelism between Environment-Context and Abstract-Concrete Norms.

4.2.4 Rules

Norms (either the *Abstract* ones or the *Concrete* ones) express constraints in a high level of abstraction, telling which are the *ideal* states of a certain environment or the acceptable behaviours in terms of quite abstract concepts such as *good, fair* or *appropriate*. Humans are able to work with such abstract concepts and, in fact, most of the *Norms* are defined in such a way that they define, for a given context, which states or actions are acceptable and which are not (we will refer to this as *the WHAT*), without defining exactly the ways to ensure or even enforce such desired states (we will refer to this as *the HOW*).

There are cases where humans are able to figure out *the HOW* by doing an interpretation directly from the normative, interpretation based in their background knowledge. For instance, most of the constraints of behaviour in the muslim orthodoxy are based on a narrow interpretation of Koranic passages. But Roman Law (and all the legal systems that are based on it) use to define explicit *regulations* for those cases where either (i) it is hard to figure out *the HOW*, or (ii) only a subset of the possible ways of action are desired. As explained in Section 3.1.2, a *Regulation* is a set of *rules* that expand a given norm defining how it may be applied. So, *the HOW* is defined by means of Rules. We can then define *Rules* as constraints which ensure that the desired states are kept or achieved, and the acceptable behaviours are performed, avoiding the rest. While norms only define in a given context C_a the boundaries of the set L_a (which are the accepted worlds), rules define the actions that are allowed inside L_a.

Definition 4.10. R_a *is the set of rules that implement the norms in* CN_a. *This can be expressed by the formula* $R_a = \text{TRANSLATION}(CN_a)$.

Where TRANSLATION: $CNorms \rightarrow Rules$ is a function that translates norms into rules, and $Rules$ is the language used to express those rules.[13]

By defining rules the normative system aims to reduce all possible interpretations of the norms to those that are closer to, what in law terminology is called the *spirit of the norm*. As rules are more concrete than Abstract and Concrete norms, the set of possible interpretations is narrowed, so it is easier for the entities (in our model the software agents) to follow them. Another clear example is the following: one of the human norms that we are told in our childhood is *"do the right thing"*. But this is too abstract to be

---

[13] We further explain how this translation is made in Section 5.4 and Section 6.4.1.

easily followed. So our parents and/or tutors give us a set of concrete rules that are to be followed in certain contexts: *"don't break items"*, *"don't shout"*, *"follow all your parents' commands"*.

In our model, the interpretation of the set of rules R_a in terms of acceptable behaviours and states inside the context C_a defines the set of *legally accessible worlds*.

Definition 4.11. $L_a \subseteq C_a$ *is the set of legally accessible worlds in context* C_a *and it is defined by the following formula:* $L_a = \text{MAP\_R}(R_a, C_a)$.

As we did in Section 4.2.1, in order to formalize the relation among the set of rules R_a and the worlds in C_a, a mapping function $\text{MAP\_R}: Rules \rightarrow 2^W$ is defined. This function is different from the MAP one, as each context may define this function in a different way.

4.2.5 Roles

The last elements in our model to define are *roles*. In previous sections the term *role* has appeared regularly. One thing to remark is that roles appear not only in one level, but in several levels, from the level of abstract norms to the level of concrete norms to the level of rules. An example in the European kingdom scenario is the role of *duke*. At the level of abstract norms there is an abstract *commander* role that then is specialized in the *duke* by means of concrete norms. Finally, at the rules' level the rights and duties of the *duke* role can be fully specified.

As this chapter is devoted to the norms from the agent view, we will only specify now the roles at the rules' level, a level that can be directly interpreted by agents. In Section 5.7 we will describe how roles are defined in the more abstract levels.

Definition 4.12. $role_n$ *is the subset of rules in* R_a *that is associated to a certain role* $role_n$.

With this definition of the set $role_n$ we can now give a more accurate definition of legally accessible worlds for a given agent a_i:

Definition 4.13. $N_i \subseteq L_a$ *is the subset of legally accessible worlds in context* C_a *that agent* a_i *can legally reach according to the rules defined in* $role_n \subseteq R_a$. *It can be defined by the following formula:* $N_i = \text{MAP\_R}(role_n, C_a)$.

Now we have a complete definition of the model from the agents' point of view. Figure 4.9 depicts the relation among the elements: agent a_i is situated in a context C_i that belongs to the world W. The normative framework has three levels of abstraction, from the abstract norms AN_w to the concrete norms CN_a associated to context C_a to the rules R_a. In such normative framework there are defined several roles, being one of them the $role_n$ played by a_i.

With all the elements presented, we can now see how the framework tries to find the equilibrium among control and autonomy presented in Chapter 1. From the society point of view, the desirable situation is that all agents follow the norms exactly as the designer intended while designing the normative framework. From the agents perspective

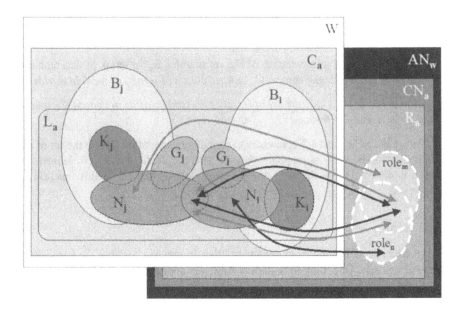

Figure 4.10: Knowledge of agent a_i and agent a_j of the rules that apply to them.

the desirable situation is to give as much autonomy to the agents to interpret the norms as possible, but this may lead to undesired emerging behaviours in the society. In our approach the organization can partially fix some of the interpretation of the norms, by defining:

- the I and TRANSLATION functions, which fix, inside the normative framework, the interpretation made from abstract to concrete norms and then to rules; [14]

- the MAP and MAP_R functions, which fix an interpretation of norms and rules in terms of possible worlds. These interpretations define only the boundaries that agents should not cross.[15]

The agents in the system keep the freedom to decide (interpret) HOW to follow the norms, and even can cross the boundaries, knowing that then a violation mechanism may be triggered. Figure 4.10 extends the example previously presented in figure 4.6 where two agents a_i and a_j are in the same context. Even though they still keep autonomy to decide how to follow norms, now the set of possible interpretations is narrower, making more likely that the agents will have similar interpretations of the same set of rules.

---

[14] see Section 5.4 and Section 6.4.1.

[15] We present how this mapping is done in terms of (verifiable) states of the *e*-organization in Section 5.4 and Section 6.4.3.

4.3   Norms and the BDI Cycle

Now that we have defined how norms have an effect in the agent from the epistemic point of view, let us now analyse norms from the operational point of view, seeing the effects of norms in the reasoning cycle of the agent. To do so we will use as basis the abstract interpreter for a BDI agent that Rao and Georgeff presented in [177]. The interpreter describes the control of a BDI agent by means of a processing cycle, from perception to action. There are several versions of the cycle, but basically it can be expressed as follows:

```
B := B_init;
I := I_init;
while (true)
{
    get_perception(perc);
    B := belief_revision(B,perc);
    D := options(B,I);
    I := filter(B,D,I);
    plan = generate_plan(B,I);
    execute(plan);
}
```

The cycle is an infinite loop where the agent starts perceiving its environment (get_perception function) and with such perceptions modifies its beliefs B about the state of the environment (belief_revision function). Then the options function generates a set D of possible alternatives (based on its beliefs and its current intentions). According to the resulting set of options D of the previous function the agent should choose one of them (filter function). The result is the current intention I the agent is committed to achieve. According to that intention, the agent creates the plan to achieve I by means of some kind of *means-end reasoning*. Finally, the plan plan is executed (execute function).

This version of the interpreter is a useful abstraction of the theoretical model of Rao and Georgeff. However this version is too simple and cannot be applied in real-time systems, as the agent does not perceive the world until it has executed the whole plan. With this first version it may happen that the state of the environment changes in a way that the execution of the plan will not have the result the agent expected, or even it may be impractical. In [178] Rao and Georgeff proposed some changes in the architecture, in representation and added some restrictions (such as having a set of pre-defined plans with invocation conditions instead of a plan generator).

The updated version of the interpreter is the following:

```
B := B_init;
I := I_init;
while (true)
{
    get_perception(perc);
    B := belief_revision(B,perc);
    D := options(B,I);
    I := filter(B,D,I);
    plan = find_plan(B,I);
```

```
while not( empty(plan) OR succeeded(I,B)
           OR impossible(I,B) )
{
   action = next_action(plan);
   execute(action);
   get_perception(perc);
   B :=belief_revision(B,perc);
   if reconsider(I,B) then
   {
       D := options(B,I);
       I := filter(B,D,I);
   }
   if not(sound(plan,I,B)) then
   {
       plan = find_plan(B,I);
   }
}
}
```

In this version most of the drawbacks of the original version have been removed. For instance, in order to avoid *overcommitment* (trying to achieve an intention that is no longer possible or trying to execute a plan that is not valid):

- plans are not executed from beginning to end but one action at a time, checking if the plan is still sound (sound function) in the current state of the world.

- intentions are checked and reconsidered at some point in time. As intention reconsideration is a quite complex process where the consistency of current intention is checked, this process is not done for each action: it is the reconsider function (usually a random function) the one that tells the agent when to check if the current intention is still valid.

As we mentioned in Section 1.3.3, one of the effects of the norms is that they reduce the set of possible actions the agent can choose. So, in order to link this interpreter with the model we presented in Section 4.1, we have to identify at which point of the reasoning cycle we must include the rules (defined in the set $role_n$).

In Section 4.1.2 we saw how in our model the goal set G_i is limited by the set N_i of *legal accessible worlds*, as the agent will enter in violations when trying to achieve the worlds in G_i that are not in N_i. Therefore, the restrictions of the normative framework have an effect on the desires and intentions of the agent. In the case of the interpreter's cycle, that means that the rules should be introduced in the intention formation steps, as the agent must, in most cases, check that the intention that it commits to achieve is legal according to the normative framework. There are two ways to do so:

- by modifying the options function to allow that only legal options are suggested. With this modification not only *feasible* options (from the point of view of the agent's capabilities) are created but only those that are *legal*.

- to modify the filter function in order to ensure that the final intention the agent chooses is a legal one.

In both cases we add an *allowance* criteria to the *feasibility* criteria already present into the interpreter.

However, as F. Dignum *et al* explain in [57], reasoning only in terms of the *allowance* criteria is not enough. Normative frameworks use to have not only restrictive norms that forbid some actions to be taken (such as *"don't kill a human being"*), but also those that impose states or actions to be taken into account into the agent's reasoning cycle (such as *"you are obliged to pay the goods you bid for"* in an Electronic Auction House).

The BDI interpreter presented in [57] adds the norms that impose actions as *deontic events* to be handled by the agent. An example is an agent A having a conditional obligation to B, $O_{AB}(\phi \mid \psi)$: when the precondition ψ becomes true, then a deontic event $O_{AB}(\phi)$ is created. Such deontic events are compared then by the agent with its goals and desires (seen also as events) and then decides which event will try to handle.

This idea can be applied to the extended BDI interpreter presented in this section by adapting the `option` and `filter` functions:

```
B := B_init;
I := I_init;
while (true)
{
    get_perception(perc);
    B := belief_revision(B,perc);
    D := options(B,I,oblEvents);
    I := filter(B,D,I,oblRestr);
    plan = find_plan(B,I);
    while not( empty(plan) OR succeeded(I,B)
               OR impossible(I,B)
    {
        action = next_action(plan);
        execute(action);
        get_perception(perc);
        B :=belief_revision(B,perc);
        if reconsider(I,B,oblEvents) then
        {
            D := options(B,I,oblEvents);
            I := filter(B,D,I,oblRestr);
        }
        if not(sound(plan,I,B)) then
        {
            plan = find_plan(B,I);
        }
    }
}
```

- the `options(B,I,oblEvents)` function now not only consider beliefs (`B`) and intentions (`I`) but also the obligation events (`oblEvents`) to be handled in the decision making,

- the `filter(B,D,I,oblRestr)` function includes now the restrictive norms (`oblRestr`) as input in order to ensure that the final intention the agent chooses is a legal one.

- the `reconsider(I,B,oblEvents)` function also should include the obligation events (`oblEvents`). With this change the agent, while executing a plan to fulfill a given intention I, might reconsider its intention because the appearance of a new obligation event that is more important than the current intention.

4.4 Enforcing and Following Norms. Police Agents

As we saw in Section 4.2.1, each agent has its set N_i of norms which defines which worlds are *legal* or *illegal* for the agent, depending on the roles it may enact. We also saw that the set $\{G_i \cap N_i\}$ defines which are the goals that the agent can *legally reach*. However in our model we allow agents to break the norms in some situations. So it is not granted that all the agents will follow all the norms anytime.

Therefore, some kind of (weak) norm enforcement is needed. In our model norm enforcement is based in the concepts of *violation* and *sanction*. As we mentioned in Section 4.2.1, a violation is a situation where a given agent breaks one or more rules, entering in an illegal state. In order to define the consequence of a violation, sanctions are also defined. A *sanction* can be defined as an action or set of actions whose realization will remove the violation. So sanctions are ways to make agents become *legal* again.

As norm enforcement is not granted, some control mechanisms should be created. The easiest way is to have a centralized controller such as the *Coordinator Agent* we mentioned in Section 1.2. In such scenario the goals of all the agents in the system are more or less defined by the tasks the *Coordinator Agent* assigns to the agents. In this scenario the Coordinator agent can, directly or indirectly change the other agents' goals in order to ensure that none of them violates the norms. However, the limitations of this approach are too serious to be acceptable (mainly *scalability* and *adaptability* issues, see Section 1.2).

So it may seem that the enforcement of norms should be completely distributed through all the agents in the system. In this scenario all agents should be aware not only of their rights and obligations, but also about the others' rights and obligations, so when there is an agent that is breaking a norm, the affected agents may detect it and punish the agent's new goal in some way. To do so, the agents should have not only a complete knowledge of the norms that apply to them but also a part of the norms that apply to other agents, and continually reason about the legacy of their own behaviour and the neighbours' one. In our model, this option means that an agent a_i must not only have knowledge about the set of rules associated to $role_n$ (that he must follow according to its role) and the interpretation N_i of such rules, but it has to know about all the sets $role_k$ defined in R_a for the other agents and their corresponding interpretations N_l in C_a. As a result, the agents may expend too many resources (i) checking other agents' behaviour, and (ii) reasoning about the roles such agents should play. Also, it is unclear which agents each agent should check, so it may happen that an agent is checked by more than one agent at the same time while there may be an agent whose behaviour is checked by no one.

F. Dignum proposes in [55] an optimization of this scenario. In his proposal the agents do not have to be aware of all the norms but only:

- the social norms that affect the agent

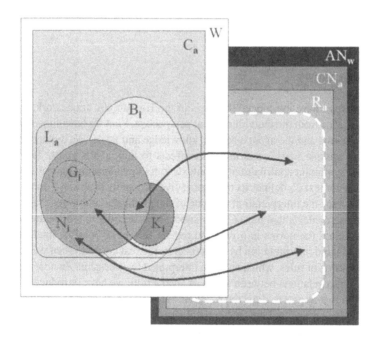

Figure 4.11: Knowledge and beliefs about norms by the Police Agent.

- the contracts the agent committed with other agent, and the countermeasures that can be taken if such contracts are violated (for instance, to ask for a reward/compensation).

The knowledge about the countermeasures to be taken in case of violation is expressed in terms of authorizations[16]. F. Dignum's proposal is sound, but it makes the assumption that all agents in the system are able to reason about obligations and authorizations, and that all them implement the extension of the BDI architecture he proposes.

A solution must be found that makes no assumption of the internal structure of the agent. A good option is the idea of *Guardian Agents* presented by Fox and Das in [75]. In our model this idea is included by defining an institutional role to enforce norms. In this scenario there is a prominent role, the *Police Agent*, that can be enacted by one or more agents. Such agents cannot access the internal code of the agents, but only perceive their actions (they see the other agents as black boxes performing actions in an environment). The Police Agent checks if the behaviour of those agents follows all the norms, (like a policeman checks the behavior of car drivers). In this case only the agents performing the Police Agent role should have a complete Knowledge of the rule set R_a, as depicted in figure 4.11.

---

[16] A review of F. Dignum's proposal can be found in Section 3.2.2.

Definition 4.14. *The* Police Agent *of a context* C_a *is a special agent such that* $role_n = R_a$ *and* $N_i = L_a$.

4.5 Summary

In this chapter we have introduced the basis of the HARMON*IA* framework, focusing on the study of norms from the individual agent perspective. First we have shown the relation between the norms and the agent's beliefs, knowledge and goals. We defined the concept of *legally acceptable worlds* and how this affects to the agent's goals. Then we have presented the important contribution of contexts to fix the interpretations of the norms. In our view each context C_i defines its own normative system by means of a consistent set of norms which fixes the interpretation of some of the predicates and terms that are *vaguely* defined in those contexts that have C_i as a subcontext. In order to reduce the abstraction, we then translated the norms into rules, to reduce the set of possible interpretations to those hat are shared interpretation by all entities inside that context. Then we have refined the norms in terms of rules, which are then interpreted by the agents. In order to represent all these complex relations between levels of abstraction, we have introduced a graphical representation that allows to draw the relations between the normative dimension defined by the normative system and the epistemic dimension of the agent.

We also have presented accurate definitions of some of the elements that compose our framework (*abstract norms, concrete norms, rules, roles* and *context*). The proposed terminology attempts to unify, as much as possible, the different contributions in the study and modelling of normative frameworks.

Then we have analysed the effect of norms in the standard BDI cycle. We have identified the extensions needed in the BDI interpreter in order to model not only the *restrictive* aspect of norms (in *prohibitions* and *permissions*) but also the *prescriptive* nature of norms, (that is, how norms expressing *obligations* induce new behaviours).

Finally, we have presented the concept of *Police Agents*, by extending the concept of Guardian Agents proposed by Fox and Das [75] in a normative scenario. The main idea is that enforcement of norms is not made through a direct control of a central authority over the goals of the agents, but through the detection of the *violation states* that agents may enter into. While the former would suppose having an access to the mental states of the agent or being able to impose goals to agents, the latter only observes the behaviour of the agent. In doing so we are able to create enforcing mechanisms for *e*-organizations which are independent from the internal architecture of the agents, a basic condition to work in open scenarios with heterogeneous agents.

The framework presented here is completed in Chapter 5, where we will describe the multi-level framework from the *e*-organization perspective. We will see how abstract norms, concrete norms and rules are placed in different levels of abstraction in the specification of the *e*-organization, how these abstraction levels are interconnected and how the surrounding context affects the *e*-organization at different levels. We will also see how violations can be defined from the rules and how to connect the rules with the final implementation of the system.

Chapter 5

A Multi-Layered Framework for Organizations

In this chapter we extend the HARMON*IA* framework presented in Section 4.2, describing the differences in levels of abstraction and defining the relations between levels from the *e*-organization perspective. Therefore, in this chapter we will take the society's point of view concerning the role of norms in the interaction between agents. Our aim is to address the following issue: how norms should be considered and included in the design and implementation process of an electronic organization (*e-Organization*). To do so we describe the process to build an agent-mediated normative electronic organization, from the statutes of the organization to the norms to the final implementation of those norms in rules and procedures. To clarify our proposal, in this chapter we will present some small examples. Then in Chapter 6 an extensive example, applying the HARMON*IA* framework to the human transplant domain.

As we have mentioned in Section 3.4.3, we will be very careful in the use of terminology. Therefore, we will abandon the widely-used term *electronic institution* to talk about *electronic organizations* or *e-organizations* to denote the multi-agent system that follows a given institutional framework, as institutions are only abstract patterns to be followed.

5.1 Creating E-Organizations

5.1.1 Using Norms inside E-Organizations

As we saw in Section 3.2 and Section 3.3, current approaches to the use of norms in the agent community either work at a low level (e.g., policies and procedures) or at a very high level, formally specifying norms using, for instance, deontic logic. The low level approaches allow an easy implementation, but the problem arises when the correctness of the procedures and policies should be checked against the original regulations. High level approaches are closer to the way regulations are made, so verification is quite easy to be done by means of model checkers. However, high level approaches usually use one or several computationally hard logics (for instance, Deontic Logic + BDI, Deontic Logic + CTL*), so direct interpretation of the logics is not a feasible option. But translating those logics into a more operational language in a single step raises the issue of checking if translation is sound and if it follows all the desired properties.

Our proposal is to work from a higher level of abstraction towards a lower level in an iterative process linking the high and low levels. In our approach getting a proper formal definition of the norms related to a certain organization is the first step in order to formally describe the organization, but that is not enough, as the level on which the norms are specified is more abstract and/or general than the level where the processes and structure of the organization are specified. Therefore, a first *translation* of the norms is needed into a level where their impact on the organization can be described directly. This translation is domain dependent, that is, it is related to the domain of the *e*-organization and, therefore, the translation steps depend on, e.g., the ontology for that domain. For instance, in an *e*-market, the norm *to pay* when a product has been bought should be translated to state the meaning of the word *pay*, that is, which kind of payment or payments are accepted (for instance, payment with money *vs.* payment with another product or products that match the value of the product paid). But even a further translation is needed to indicate *how* the norm is implemented. For instance, the norm to perform a monetary payment when you have bought a product can be implemented by restricting the available actions in the organization after the buying action to just the paying action. However, one might also implement this norm by not allowing an agent to leave the organization before he has paid (in case he bought something). This means that the agent can still perform all kinds of actions, but always has to pay at some time.

5.1.2   Four Levels of Abstraction

The HARMON*IA* framework is a multi-level approach to model *e*-organizations. It is composed by four levels of abstraction:

- the *Abstract Level*: where the statutes of the organization are defined in a high level of abstraction along with the first abstract norms.

- the *Concrete Level*: where abstract norms are iteratively concretized into more concrete norms, and the policies of the organization are also defined.

- the *Rule Level*: where the concrete norms are fully refined, linking the norms with the ways to ensure them. The policies defined in the previous level are refined.

- the *Procedure Level*: where all rules and policies are translated in a computationally efficient implementation easy to be used by agents.

The division of an *e*-organization into these four levels aims to ease the transition from the very abstract statutes, norms and regulations to the very concrete protocols and procedures implemented in the system, filling the gap among previous theoretical (abstract) approaches and practical (concrete) ones.

The abstract and concrete levels compose the *Normative System* of the *e*-organization. They define all the *obligations*, *rights* and *permissions* that should be applied in the system. The rule and procedure levels compose the *Practical System* of the *e*-organization, as they define how the agents inside the system should behave (the *praxis*) in order to follow the norms in the normative system, and how the *e*-organization can enforce the fulfillment of, at least, some of the norms.[1]

Definition 5.1. *An* Electronic Organization *(e-Organization) is a tuple*

$$eOrg = \langle NS, PS, role\_h, policies, ontologies \rangle$$

where:

- $NS = \langle abslevel, conlevel \rangle$ *is the* Normative System, *composed by the* Abstract Level *(abslevel) and the* Concrete Level *(conlevel)*,

- $PS = \langle rulelevel, proclevel \rangle$ *is the* Practical System, *composed by the* Rule Level *(rulelevel) and the* Procedure Level *(proclevel)*,

- *role_h is the* Role Hierarchy *(see definition 5.9)*,

- *policies* $= \langle policy_1, policy_2, \ldots, policy_n \rangle$ *is the set of policies defined in the e-organization (see definition 5.8)*,

- *ontologies* $= \langle ontology_1, ontology_2, \ldots, ontology_m \rangle$ *is the set of ontologies used inside the e-Organization.*

The multi-level framework is depicted in figure 5.1. It depicts the refinement process where norms are defined, starting with the organization's statutes in the abstract level, and then progressively refined in more concrete norms in the concrete level. Then all those norms are translated into rules and, finally, implemented at the procedure level. We will describe this process in detail, level by level, in sections Section 5.2 to Section 5.5. Then in sections Section 5.6 to Section 5.8 we will present some elements that are defined through all four levels: the *policies* of the *e*-organization, the *role hierarchy* (and the goal definition attached to it) and the *ontologies* used by the *e*-organization.

5.2 The Abstract Level: Statutes, Objectives, Values and Norms

The characteristic of institutions is that they enforce certain behaviors among members of the group of persons (or agents) that carry out their interactions within the jurisdiction (scope) of the organization. These behaviors must be of a type that contribute to a given

---

[1]This distinction between normative and practical aspects is not new. The first reference is by Hume (1739 aC.), in a famous passage called *"the is-ought passage"*. The passage inspired the so called *Hume's Law* (stated by Moore): *"One cannot move from premises containing only descriptive terms or concepts to a conclusion that contains normative or evaluative terms or concepts."*

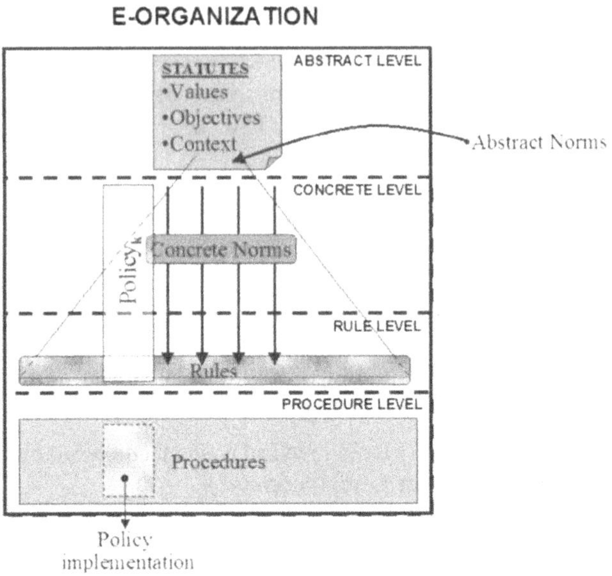

E-ORGANIZATION

Figure 5.1: Multi-level architecture for virtual organizations.

pre-specified objective and set of values and norms. At the most abstract level these three elements can be found in the *statutes* of the organization. The statutes will indicate the main *objective* of the organization, the *values* that direct the fulfilling of this objective and they also point to the *context* where the organization will have to perform its activities.

For example, the statutes of the National Organization for Transplants (ONT) [155] in Spain state the following:

> *The principal objective of the ONT is therefore the promotion of donation and the consequent increase of organs available for transplantation, from which all its other functions result. The ONT acts as a service agency for the whole National Health System, works for the continuing increase in the availability of organs and tissues for transplantation and guarantees the most appropriate and correct distribution, in accordance with the degree of technical knowledge and ethical principles of equity which should prevail in the transplant activity.*

In that statement we can find:

1. the *objectives*: the main objective of this organization is to increase the number of organ donations and the subsequent increase in available organs for transplants. Another objective is to properly allocate organs.

2. the *context*: ONT states that it operates inside the Spanish National Health System, and such statement clearly defines the context of the organization.

3. the *values*: The latter part indicates the values according to which the ONT operates. The values are the following: to guarantee the most *appropriate*, *correct* and *equal* distribution of pieces among potential recipients. Where *appropriate*, *correct* and *equal* are vague terms defined by both technical (medical) and ethical standards. Implicitly it also says that ethical values are also part of the organization. These values will also play a role in the regulations that will determine the actual process according to which the transplantation should be performed.

At this level of abstraction, the highest, *values* fulfill the role of norms in the sense that they determine the actions that individuals should or should not take in a certain situation. Values are beliefs that individuals have about *what* is important, both to them and to society as a whole. A value, therefore, is a belief (right or wrong) about the way something should be. It should be apparent that values, by definition, always involve judgements (since they tell *how* something should be). In short, the values individuals hold are general behavioral guidelines. Values define beliefs about, for instance, what is acceptable or unacceptable, good or bad. However, values do not specify *how*, *when* or in *which* conditions individuals should behave appropriately in any given social setup. This is the part played by abstract norms, concrete norms (see Section 5.3) and rules (see Section 5.4).

The *objectives* of the organization can be represented as the goal of such organization. As far as the organization has control over the actions of the agents acting within that organization, it will try to ensure that they perform actions that will lead to the overall goal of the society. See Section 5.7 for more details.

We will talk about the contribution of *context* in Section 5.9.

Definition 5.2. *The* statutes *of a given e-Organization are a tuple*

$$statutes = \langle values, objectives, context \rangle$$

where:

- $values = \langle value_1, value_2, \ldots, value_n \rangle$ *is the set of* Values *of the e-Organization (see definition 5.3),*

- $objectives = \langle objective_1, objective_2, \ldots, objective_m \rangle$ *is the set of objectives the e-Organization aims to fulfill,*

- $context = \langle eOrg_1, eOrg_2, \ldots, eOrg_l \rangle$ *is the set of electronic Organizations that influence the behaviour of the e-Organization. The set context can be void ($l = 0$).*

5.2.1 From Values to Norms

The values of the organization can be described as desires. For example:

- the value of *appropriate distribution of donated organs* (which is related to medical issues such as *histo-compatibility*), can be described as $D(appropriate(distribution))$,

- the value of *correct distribution of donated organs* (which is related to management issues such as an efficient organization of the system) can be described as $D(correct(distribution))$,

- the value of *equal distribution of donated organs* (which is related to ethical issues) can be described as $D(equal(distribution))$.

However, besides a formal syntax, this does not provide any meaning to the concept of *value*. Here, we do not intend to define a logic about values, but only use them as an initial step to derive the normative system. In our framework the meaning of the values is defined by the abstract norms that contribute to this value. In an intuitive way we can see this translation process as follows:

$$\vdash_{org} D(\varphi) \mapsto O_{org}(\varphi)$$

meaning that, if an organization org has φ as one of its values, then such value can be translated in terms of an abstract norm (an obligation of the organization org to fulfill φ). But it usually happens that the value cannot be fully expressed by means of a single norm but several ones that contribute to a value. In our framework a norm *contributes to a value* if fulfilling the norm always leads to states in which the value is more fully accomplished than the states where the norm is not fulfilled. For instance, a norm contributing to the *appropriate* value would be that *"donor/recipient compatibility should always be checked"* but also another one could be that *"the quality of the organ itself should be guaranteed"*. So, each value has attached to it a list of one or several norms that contribute to that value. The total list of norms (the ones in the abstract level plus the ones in the concrete level) together *defines* the meaning of the value in the context of the organization.

$$\vdash_{ONT} D(appropriate(distribution)) \quad \mapsto \quad \{O_{ONT}(\text{"ensure an appropriate distribution"})\}$$

$$\vdash_{ONT} D(correct(distribution)) \quad \mapsto \quad \{O_{ONT}(\text{"ensure an efficient distribution"}),$$
$$O_{ONT}(\text{"reduce the time of assignation"}),$$
$$...\}$$

Definition 5.3. *A* value *is a tuple*

$$value_i = \langle val, a\_norms \rangle$$

where:

- val *is the representation of the value in a language* $Values$,

- a_norms *is the set of abstract norms that contribute to the value.*

5.2.2 Representing Norms

Having set the most abstract level of normative behavior in the statutes, we will now turn to the description of the norms themselves in the abstract level.

To do so, we have to determine how the norms will be represented. Of course the representation of the norms will determine, in a large part, the level of abstraction and the way they can be related to more concrete norms. For instance, if we represent all obligations by means of a classical (modal) deontic logic based on prepositional logic then every state of affairs is collapsed into a proposition and relations between states of affairs might be difficult to represent.

There are different needs depending on the abstractness of the norms. *Abstract Norms* (the norms in the Abstract Level) are simpler than the *Concrete Norms* (the norms in the Concrete Level). Therefore, *Abstract Norms* can be expressed as either static or dynamic deontic logic formulæ, while *Concrete Norms* usually need dynamic deontic formulæ (to express time constraints or cause-effect relationships) and some kind of mechanism to handle context (as we saw in Section 4.2.2, contexts can be nested one into another, so the logic should handle the effects of the contexts in the normative system).

With this idea in mind, in Section 4.2.3 we define *ANorms* (the language for abstract norms) and *CNorms* (the language for concrete norms), and a function I: $ANorms \rightarrow CNorms$ which is a mapping from the abstract norms to the concrete ones. Even though our framework allows the use of different languages for abstract and concrete norms, for the examples in this chapter we will assume that $ANorms = CNorms$, in order to simplify the complexity, choosing a deontic logic that is temporal, relativized and conditional. That is, an obligation to perform an action or to reach a state can be conditional on some state of affairs to hold, it is also meant for a certain type (or role) of agents and should be fulfilled before a certain point in time.

The consequence of the above is that we assume that norms can be categorized in three types: obligations ($O_a(\varphi)$), prohibitions ($F_a(\varphi)$) and permissions ($P_a(\varphi)$). The obligations and prohibitions are mainly seen, in this context, as constraints upon the possible actions (and subsequent situations) while the permissions are generally used to indicate the conditions in which, otherwise forbidden actions, can be performed.

Using this deontic logic we can formalize the previous norm *"The ONT must ensure an appropriate distribution"* as follows:

$$O_{ONT}(appropriate(distribution))$$

The obligation is directed towards the ONT, which is responsible to ensure that the distribution is appropriate. Another example of norm to be formalized is: *"The donor should consent to the transplantation before the transplantation can take place"*. In this case such norm can be expressed through the following formula:

$$O_{hospital}(consent(donor) < do(transplant(hospital, donor, recipient, organ)))$$

The obligation is directed in this second example towards a hospital, assuming that the hospital is responsible for fulfilling it. That is, it is the responsibility of the hospital to acquire the consent of the donor before the transplantation is performed[2]. In some sense

---

[2]This example belongs to those cases where the transplantation is made with a living donor. Examples of organs and tissues that can be transplanted in this way are marrow cells, a kidney or half the liver.

this obligation has an implicit conditional. It only comes into effect if the hospital intends to perform a transplantation.

Definition 5.4. *The* Abstract Level *of an e-Organization is a tuple*

$$abslevel = \langle statutes, a\_norms, ontology_{al} \rangle$$

where:

- *statutes are the* Statutes *of the e-Organization, as described in Definition 5.2,*

- $a\_norms = \langle norm_1, norm_2, \ldots, norm_n \rangle$ *is the set of* Abstract Norms, *expressed in the language ANorms,*

- $ontology_{al}$ *is the ontology that defines the terms and predicates appearing in the norms in a_norms.*

5.3   The Concrete Level: from Abstract Norms to Concrete Norms

Norms are expected, socially acceptable, standards of behaving in any given social situation. In order to check norms and act against possible violations of the norms by the agents within an organization, the abstract norms have to be translated into actions and concepts that can be handled by the agents within such organization. Hence, *concrete norms* should refer to a) actions that are described in terms of the ontology of the organization and from which therefore the meaning and effect is known, or b) they pertain to situations that can be checked directly by the organization. For instance, a concrete norm might be: *"a buyer should pay a deposit before he can start bidding at an auction"*. In this case *pay deposit* and *bidding* are basic actions within the organization. A norm such as *"a buyer must be of legal age"*[3] might be concrete if the organization has a means to check the age of the buyer directly. However, if the organization and/or the buyer must follow some procedure to proof the age of the buyer it is considered an abstract norm.

5.3.1   Sources of Abstraction

Of course one can categorize norms in many different ways. In this section we will categorize the norms in the ways they are abstract, i.e., in the way that they generalize from or are more abstract then the concepts that are used to define the functioning of the organization. Subsequently, we will indicate how a translation can be made between the abstract norms and more concrete forms.

The abstract norms try to capture many different situations and therefore are *abstract* in several different ways:

1. They are referring to an abstract action that can be implemented in many ways

---

[3]The *legal age* concept varies among states. In western countries it ranges from 16 to 21 years

2. They use terms that are vague and that have to be defined separately

3. They abstract from temporal aspects

4. They abstract from agents and or roles

5. They refer to actions or situations that are not (directly) controllable and/or verifiable by the organization

Abstract Actions

The first category is perhaps the most intuitive one. An example is the following: *"a living donor should consent to the donation of an organ"*. The action of *consent* might not be an action that is described in terms of the concepts used in the ontology of the *e*-organization and is therefore still abstract. It can be implemented by an action of the donor signing a contract or the donor telling it to his family or the donor carrying a will, etc. Each of which might be a valid action within the organization. So, the first case of abstraction refers to an abstraction of actions such that they refer to a set of possible ways of performing the action. Each of the possible ways of performing the abstract action (defined within the organization) should be either permitted, prohibited or obliged within the organization. The translation in this case is a kind of definition of the abstract action in terms of the concrete actions. In the above case one could define this as follows:

$$sign(donor, contract) \cup carry(donor, will) \cup tell(donor, family) \Rrightarrow_{SpNHS} Consent(donor)$$

that is, performing either of the three more specific actions counts as[4] (in the context of the Spanish National Health System), *to consent*. Important to note is that this definition closes the way consent can be given. In the above case it should be one of the three ways mentioned.

Vague Terms

The second category of norms cannot be implemented directly because the norms use terms that have no precise meaning. For example, *"the extraction of organs from living donors should be limited to those situations where one can expect a high probability of success of the transplant"*. This is a very common phenomenon in legal texts. *Vague terms*[5] are often explicitly used to allow for interpretation within different contexts.

In this case it is not specified what it means to *expect a high probability of success*. There are two ways in which these terms can be explicated:

1. through a rule that defines the meaning of the term. For example, *"if the data from the donor's file and that of the recipient's file match for more than 90% and the recipient is otherwise in good condition and the donor's organ is healthy then one can expect a high probability of success"*.

---

[4]During the refinement process we use a variation of the *counts-as* connective ($\Rightarrow_s$) defined by Jones and Sergot [107]. The $\Rrightarrow_s$ connective proposed by Dignum and Grossi in [90, 89] keeps the same intuitive meaning ($A \Rrightarrow_s B =$ "*A counts as B in context s*") but modifies its definition to handle non-monotonicity.

[5]See [98] for more discussion on *vague terms*.

2. through a procedure that has to be followed in order to determine whether the condition referred to in the term is true. For example, *"One can expect a high probability of success if the medical committee of the recipient's hospital has consulted the data from the files of both donor and recipient and declared that the probability of success is higher than 90%".*

The first case actually replaces a condition that cannot be checked directly by the organization by a number of conditions that can be checked by (or within) the organization. In the second case there are no available data to be checked, but a procedure can be followed to determine the value of the condition. It also implies that the only (accepted) way to find out the value of the condition is through the given procedure. Both cases have in common that one cannot proof directly whether the condition referred to in the norm is true or not. In order to proof that the condition is true one has to check some other conditions (either directly or through some prescribed procedure).

Applying this to the example of the appropriateness of the distribution, the *appropriate* term is so vague that there is a need to refine it in terms that are easier to check and/or implement:

$$O_{ONT}(ensure\_quality(organ)) \land O_{ONT}(ensure\_compatibility(organ, recipient))$$
$$\Rightarrow_{ONT} O_{ONT}(appropriate(distribution))$$

In this case the appropriateness has been refined in terms of two procedures to be checked: 1) to check that the organ has the minimum quality required to be transplanted, and 2) to ensure that the organ and the recipient are compatible[6].

Temporal Abstractness

The next category of norms refers to norms where temporal aspects are undefined or vaguely defined. Often there is an implicit deadline for obligations, which is implied by the fact that the fulfillment of an obligation is also the fulfillment of a condition for a permission. For instance, the consent of the donor is a condition of the permission to perform a transplantation. Therefore, the consent of the donor should be given before the transplant can be carried out. Although the obligation to get the consent does not have a deadline, the rest of the procedure has to wait for this action. It might be that some deadlines are defined on other parts that indirectly limit the time to fulfill the obligation. For example, if the whole procedure of the transplantation should be finished within one week (or otherwise the recipient might die) then the consent should be given within this week.

Returning to the *appropriateness* example, there is a temporal relation that is missing there. The obligation of ensuring the quality and compatibility of the organ is limited by the temporal constraints of the assignation process –organs cannot be preserved out of a human body for more than some hours–, as the quality and compatibility tests should be

---

[6]This refinement has not been done randomly. As we will see in Section 5.9, this refinement is imposed by the context of the ONT.

performed before the assignation is made. We can then extend the formula as follows:[7]

$$O_{ONT}(ensure\_quality(organ) < do(assign(organ, recipient))) \wedge$$

$$O_{ONT}(ensure\_compatibility(organ, recipient) < do(assign(organ, recipient)))$$

Agent and Role Abstraction

The fourth category of abstraction is about the role(s) or agent(s) for whom the norm holds. There is often an implicit assumption about who is the responsible for an action or situation. Therefore, the obligations, prohibitions and permissions are stated in a general form. They are supposed to affect any agent that wants to perform a transaction through the *e*-organization and which is eligible to do so according to the regulations for admission of the *e*-organization. For example, tissues can only be preserved in tissue banks This process is done as the roles are refined in the *role hierarchy* (see Section 5.7).

Actions or Situations not Directly Verifiable

The last category of norms refers to actions and situations that are not directly verifiable. An example is the following: *"the decision of who is the best recipient for an organ cannot be based on the age of the recipient"*. Although the norm is clear, it is impossible to check directly on which basis a decision is taken by an agent. This is an internal (mental) action. In order to check this norm, the organization has to devise some constraints and/or procedures that are verifiable (or controllable) by the organization and which take care of the fulfillment of the norm. For instance, the *e*-organization might withhold all information pertaining to the race of the potential recipients to the decision makers. Alternatively, they might let the decision makers sign a contract in which they promise not to let the race of the potential recipients influence their decisions.

The same type of argument holds for situations that are not directly verifiable by the organization. For instance, *"the donor should be of legal age"*. The fact is that in order for the organization to uphold this norm, it should be able to know whether the state described in it holds or not. Knowing in this context means that the organization has some kind of proof which it deems sufficient in order to claim it knows the state. Thus, the organization has to have a procedure or another fact which can count as evidence for the state of affairs. This is very similar to the situation in commerce. If the contract states that the buyer should pay as soon as the seller shipped the product then the question becomes how the buyer can make certain that the seller actually shipped the product before he pays. Also, in this case there are a number of principles that can be applied to check whether the procedure that is followed actually assures that both parties have enough evidence (knowledge) to act upon.

---

[7]Note that we have introduced a boolean predicate *do* that, applied to an action, it becomes true when the action starts to be performed. This is needed in the Temporal Deontic Logic we use here (that is presented in Section 3.2.1), an expression such as $O(p < q)$ means that action p should be performed before condition q holds.

Definition 5.5. *The* Concrete Level *of an e-Organization is a tuple*

$$conlevel = \langle c\_norms, ontology_{cl} \rangle$$

where:

- $c\_norms = \langle norm_1, norm_2, \ldots, norm_m \rangle$ *is the set of* Concrete Norms, *expressed in the language $CNorms$,*

- $ontology_{cl}$ *is the ontology that defines the terms and predicates appearing in the norms in c_norms.*

5.3.2   Limitations of Norms

Given the above classification we can define a formal language $CNorms$ in which we translate the more abstract norms into the concrete ones (like was illustrated in the previous example). However, even the concrete norms have no direct implementation in an organization. The obligation to sign a donor contract before a transplantation takes place can be formally represented as:

$$O\big(signed(donor,\ contract) < do(transplant(hospital,donor,\ recipient,\ organ))\big)\big)$$

that is, in the situation a transplant is taking place, it is obliged to have a contract signed by the donor. However, this obligation does not state how this signed contract comes about. Therefore an organization cannot be fully modelled by means of norms. This is because norms work only in a normative level, which defines what is acceptable (legal) and what is unacceptable (illegal), but there is also a need to define the actions to carry out (e.g., the sanctions) in the descriptive level.

    Sanctions are a good example to remark the difference among the normative and the descriptive levels. Even though in a given Law we may find, in the same document, norms and sanctions, the sanctions are ways to *implement the norms* in that document. Such a distinction should be made among the norms (that are normative) and the sanctions, rules and other implementations of norms (that are descriptive).

    With this distinction in mind it is clear that an organization cannot be fully modelled by means of norms, as we also need to describe, at least, some of the behaviours and courses of action that can be followed inside the *e*-organization.

5.4   The Rule Level: Translating Norms into Rules

The translation from norms to rules marks the border between the *Normative System* in HARMON*IA* and the *Practical System*, from the normative dimension to the descriptive one. Such translation also implies a change in the language, from a deontic logic to a language more suitable to express actions and time constraints.

Meyer proposed in [138] a reduction from deontic logic to a Propositional Dynamic Logic, that is based in the modal operator [ ]. In his approach, deontic formulæ such as $O(\alpha)$, $F(\alpha)$ and $P(\alpha)$ are reduced to dynamic logic as follows:

$$F(\alpha) \mapsto [\alpha] \, V$$

This formula is the basis of our translation from norms to rules. Informally, it says that the action α is forbidden if and only if the performance of α leads necessarily to a state in which the violation V holds. It can also be interpreted in the opposite direction: if the action α is forbidden, then the execution of α triggers the violation V.

Meyer, from this translation rule, also defines the translations for obligations and permissions:

$$P(\alpha) \equiv \neg F(\alpha) \mapsto\, <\alpha> \neg V$$

$$O(\alpha) \equiv F(\neg \alpha) \mapsto [\neg \alpha] \, V$$

Informally, the first says that α is permitted if and only if α is not forbidden, so there is a way to perform α that leads to a state where violation V is not triggered. The second one expresses that α is obligatory if and only if not doing α is forbidden (and then it will trigger a violation).

In those cases where the norm expresses temporal relations among predicates, such relation can be translated to Dynamic Logic as follows:

$$O(\alpha < do(\beta)) \mapsto [\beta] \, done(\alpha)$$
$$O(\alpha < do(\beta)) \mapsto \neg done(\alpha) \rightarrow [\beta] \, V$$

The first reduction rule translates the temporal constraint of α being done before β starts with an expression in Dynamic Logic that states: *"once action β is performed, it should always be the case that action α has been done"*. The second reduction rule expresses the violation condition: *"if action α has not been done, once action β is performed it always is the case that violation V occurs"*. Therefore, concrete norms can be translated into two kinds of rules: *precedence rules* (rules defining precedence between actions) and *violation rules* (rules defining condition where a given norm is violated).

Examples of this kind of translation in the organ and tissue allocation problem are the following:

$$O_{hospital}(ensure\_quality(organ)) \mapsto [\neg ensure\_quality(organ)] \, V(hospital)$$

$$O_{ONT}(ensure\_compatibility(organ, recipient) < do(assign(organ, recipient)))$$
$$\mapsto [assign(organ, recipient))] done(ensure\_compatibility(organ, recipient))$$

From the point of view of the organization one would like to create a situation in which it is impossible for the agents to violate any of the specified norms, because in that situation the organization could guarantee that all the transactions are performed according to the objectives and values indicated in its statutes.

- In the case of *prohibitions*, the enforcement consists on (trying to) make it impossible for the agents to perform an action that it is not allowed to perform. For instance, if an agent is not registered at an auction it is not allowed to bid at the auction. The latter can be enforced by just preventing an agent to enter the auction before it is registered.

- In the case of *obligations*, as norms indicate that some behavior is wanted in certain situations (e.g., paying after buying something), they are translated as much as possible as constraints on unwanted behavior. The difference is that the *e*-organization cannot actually force an agent to pay (this is an agent's autonomous decision), but it can control the fact that the agent cannot leave before it has paid.

- In the case of *permissions*, norms indicate the allowance of actions under certain conditions. These norms can be translated to checks on the conditions whenever an agent tries to perform the action. If the conditions are not fulfilled, the agent is prevented from performing the action (if possible).

Therefore, following this idea, norms can be translated some part into restrictions on behavior and the other part into triggers on unwanted behavior of the agents interacting in the *e*-organization. By translating as many norms as possible into behaviors that can be controlled by the organization, the organization can enforce as many norms as possible.

The concept of *violations* is a very important one, as violations allow the *e*-organization to keep an acceptable control of the behaviour of the agents if some actions of the agents are uncontrollable by the organization or if controlling all of them is a very time-consuming task:

- In the case that there are actions which are not under the (direct) control of the organization, violations are the only way to try to avoid agents inside the *e*-organization to choose those actions. For instance, the bidders in an auction are not allowed to make agreements or to communicate outside the auction about price arrangements. However, the institution cannot really enforce this prohibition, it can only trigger a violation if it detects such behaviour. Another example is the *policeman vs. car driver* one presented in Section 4.2.1, where the policeman cannot physically prevent the car driver exceeding the speed limit but only fine him if it detects the violation of the norm.

- There are cases where trying to control all possible behaviors to enforce normative behavior can result in a very inefficient or cumbersome functioning of the organization. Therefore, in some cases the organization might decide, for the sake of flexibility and efficiency, to not enforce a norm through restricting behavior but react on violations of norms.

Once the translation from norms to rules has been done, some refinement can be done to those rules by adding more rules. One example of this refinement process at the rule level is to refine the $ensure\_compatibility(organ, recipient))$ predicate by means of donor-recipient compatibility rules. For instance, in the case of the kidney, those rules are the following[8]:

```
1- (age_donor <= 1)
      -> (age_recipient < 2)

2- (age_donor > 1) AND (age_donor < 4)
      -> (age_recipient < 4)

3- (age_donor >= 4) AND (age_donor < 12)
      -> (age_recipient > 4) AND (age_recipient < 60)

4- (age_donor >= 12) AND (age_donor < 60)
      -> (age_recipient >= 12) AND (age_recipient < 60)

5- (age_donor >= 60) AND (age_donor < 74) AND (creatinine_clearance > 55 ml/min)
      -> (age_recipient >= 60) AND (transplant_type SINGLE-KIDNEY)

6- (age_donor >= 60) AND (age_donor < 74) AND (glomerulosclerosis <= 15%)
      -> (age_recipient >= 60) AND (transplant_type SINGLE-KIDNEY)

7- (age_donor >= 60) AND (glomerulosclerosis > 15%) AND (glomerulosclerosis <= 30%)
      -> (age_recipient >= 60) AND (transplant_type DUAL-KIDNEYS)

8- (weight_donor = X)
      -> (weight_recipient > X*0.8) AND
         (weight_recipient < X*1.2)

9- (disease_donor Hepatitis_B)
      -> (disease_recipient Hepatitis_B)

10-(disease_donor Hepatitis_C)
      -> (disease_recipient Hepatitis_C)
```

Another example is to refine the $ensure\_quality(organ)$ for the case of kidneys:

```
11-(disease_donor HIV)
      -> (DISCARD-DONOR)

12-(glomerulosclerosis > 30%)
      -> (DISCARD-KIDNEY)

13-(HLA_compatibility_factors < 3)
      -> (DONOR-RECIPIENT-INCOMPATIBILITY)
```

In both cases, these refinements are driven by the standards defined by the context, as we will see in Section 5.9.

Definition 5.6. *The* Rule Level *of an e-Organization is a tuple*

$$rulelevel = \langle rules, violations, ontology_{rl} \rangle$$

where:

- *$rules = \langle rule_1, rule_2, \ldots, rule_l \rangle$ is the set of* Rules, *expressed in the language* Rules,

- *$violations = \langle viol_1, viol_2, \ldots, viol_l \rangle$ is the set of* Violations *to be checked,*

- *$ontology_{rl}$ is the ontology that defines the terms and predicates appearing in the rules set.*

---

[8]These rules were already presented in Section 2.5.

5.5 The Procedure Level

The final step to build the *e*-organization is to implement the multi-agent system. There are two main approaches to be followed:

- The implementation of the rules in the rule level is done by using rule interpreters. This means that any agent entering the *e*-organization will need a suitable rule interpreter in order to behave properly. This will allow agents to read and reason directly about the rules.

- The implementation of the rules in the rule level is done by translating the rules into procedures to be easily followed by the agents. In this case the agents will be more efficient as they simply have to follow protocols described in a procedural way instead of reasoning about the rules. In the case of rules expressed in Dynamic Logic, this translation not only is easy to be done but also is principled by the following expression connecting Dynamic Logic formulæ with Hoare formulæ:

$$\{P\}\alpha\{Q\} \equiv P \to [\alpha]Q$$

In both cases, representing the norms by rules, procedures or any other description does not ensure that the agents will follow those descriptions. Also, the violations in the rule level should be translated in some detection mechanisms to check the behaviour of the agents.

The two approaches (interpreting rules / following protocols) are not mutually exclusive. A system where all external agents blindly follow the protocols (as in ISLANDER [68]) is quite efficient from a computational perspective but the agents have only the autonomy to accept or reject the protocol, and when they accept to follow it then they loose their autonomy and their proactiveness (for instance, they cannot anticipate and react to situations that the protocol did not foresee, and thus the agents may be performing the wrong actions, or those that not make them closer to achieve their goals). A system where all external agents should be able to interpret rules could cope with these problems, as the actions of the agents are not tightly limited by state protocols but to rules that leave the agent with more options to choose. However, such a system could not be applied to open environments, as there is a big assumption in part of the internal architecture of the agents and their reasoning methods.

The alternative is to be able to accept both kinds of agents. That means that the *e*-organization provides, to all external agents entering into the organization, with the low-level protocols and also with the related rules. Those (standard) *Autonomous Agents* that are only able to follow protocols[9], will blindly follow them, while the ones that can also interpret the rules (that we call *Flexible Normative Agents*) will be able to choose among following the protocol or reasoning about the rules, or do both. With this approach the autonomy of the agents entering the *e*-organization is adapted to their reasoning capabilities.

---

[9]These protocols could be expressed, for instance, in DAML+OIL, language to describe services that is currently used in most projects of the *Agentcities* initiative.

The *Flexible Normative Agents* –that we propose here– are a special kind of *Norm Autonomous Agents* (presented in Section 1.1 and Section 3.2.2) where the agents can not only interpret and reason about rules and the related norms but also, in those cases where time efficiency is needed, follow a protocol. For instance, a *Flexible Normative Agent* will usually follow the protocol -that is the computationally less expensive option- but it will also regularly check the rules to identify those cases that were unforeseen by the protocol or those that their behaviour does not meet its expectations, by choosing another course of action).

In order to allow *Flexible Normative Agents* to switch from following low-level protocols to higher level rules and norms, there should be a link from procedures to rules, and from rules to norms. An advantage of the HARMONIA framework is that those links are created by the designer in the process from abstract to concrete norms to rules by means of the successive translations that are made. Those translations are links that allow to track, for instance, which abstract norms are related to a given rule. So the same should be made in the translation from rules to procedures. Figure 5.2 gives an example of such linking.

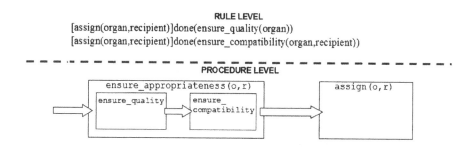

RULE LEVEL
[assign(organ,recipient)]done(ensure_quality(organ))
[assign(organ,recipient)]done(ensure_compatibility(organ,recipient))

PROCEDURE LEVEL

ensure_appropriateness(o,r)
ensure_quality   ensure_compatibility

assign(o,r)

Figure 5.2: Translation from rules to procedures.

The rule in the rule level only states that the quality and compatibility of the organ should be ensured before assigning it to a given recipient. The predicates of the rule are then translated into procedures that are part of the assignation protocol, and a link from the predicates to the procedures is kept. It is important to note here that the implementation of the rule is more restrictive than the rule (as it imposes the quality assessment to be done *before* the compatibility assessment), but such restriction is not present in the original rule. It is only an implementation decision taken by the designer, which could also choose to do both tests in parallel, or in reverse order. A standard agent that blindly follows the protocol will always wait for the ensure_quality procedure to end before executing the ensure_compatibility one, even in the extreme situations where there is some technical problem that temporarily makes the quality assessment to be able to return results while the compatibility assessment process is running. On the other hand, a *Flexible Normative Agent* will be able to detect a unusual delay or wrong performance

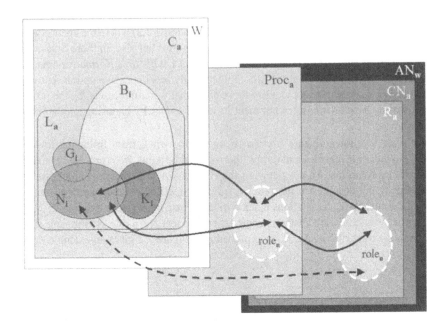

Figure 5.3: The Procedure Level from the agent's point of view.

of the `ensure_quality` procedure, see that the `ensure_compatibility` one is running, reason that there is a time constraint to perform the assignation and then decide to check if it is legal to do the compatibility assessment instead (which, in this case is allowed by the rule, as no violation will be triggered by doing the compatibility assessment first). Hence, in this case the agent will stop following the protocol to cope with the abnormal situation, while keeping its behaviour inside *legality*.

From the agent's perspective (presented in Section 4.2) this duality can be represented by adding the Procedure Level in the diagram. Figure 5.3 depicts this new level, and how the agent's perception of the norms can come from an interpretation of the procedures in the procedure level or, optionally, by directly interpreting the rules in the rule level.

Definition 5.7. *The* Procedure Level *of an e-Organization is a tuple*

$$proclevel = \langle procedures, plans, protocols, ontology_{pl} \rangle$$

where:

- *procedures* $= \langle proc_1, proc_2, \ldots, proc_n \rangle$ *is the set of pre-defined* Procedures *in the system, including the triggers and sanctions mechanisms to enforce norms,*

- *plan* $= \langle plan_1, plan_2, \ldots, plan_n \rangle$ *is the set of pre-defined* Plans *in the system,*

- $protocols = \langle prtl_1, prtl_2, \ldots, prtl_m \rangle$ *is the set of communication protocols to be used in the e-Organization,*

- $ontology_{pl}$ *is the set of ontologies (including the communication ontologies) used by the agents in the e-Organization.*

5.6 Policies

Policies are usually thought to be an important element of organizations. The term *policy* is one of the most confusing ones in the literature, though. Some authors identify policies as plans, some others as protocols, or to procedures. The reason that leads to this confusing situation is that policies are both abstract and concrete. The Webster Dictionary defines *policy* as follows:

2a : a definite course or method of action selected from among alternatives and in the light of given conditions to guide and determine present and future decisions

2b : a high-level overall plan embracing the general goals and acceptable procedures especially of a governmental body

Definition *2a* defines policies as low-level, procedural elements that affect the decision making process (a policy implementation), while *2b* defines *policies* as high-level plans that relate the generic goals of an organization with the procedures. With this in mind, in our framework policies are vertical elements that, following definition *2b*, go from the abstract level (where goals and values are defined) to the rule level (where those values are described in terms of rules). Following definition *2a*, we will call *policy implementation* the implementation of the norms and rules of the policy in the procedural level. Figure 5.1 depicts policies as columns linking the abstract, concrete and rule levels, which have an associated policy implementation in the procedure level.

Let us point out some examples in the organ and tissue allocation problem. An *e*-organization trying to fulfill the role of the ONT will have to define, at least:

- an *Allocation policy*, as the values and objectives of ONT state that it aims to perform a *correct* and *appropriate* allocation. This policy starts in the statutes of the ONT (as values and objectives to obtain), then is iteratively refined by means of norms and rules (such as the ongoing appropriateness example) and then is implemented in the procedure level.

- a *Security policy* (imposed by the context, as we will see in Section 5.9) for all the information about patients it manages. In this case it starts from an objective to keep the anonymity of the patients by avoiding unauthorized entities to access to their information, then it will be refined in terms of norms and rules. The implementation of the policy consists of all the security mechanisms implemented in the multi-agent system to ensure privacy.

An extensive example of policies, defined from the abstract level to the procedure level, can be found in Chapter 6.

Definition 5.8. *A* Policy *of an e-Organization is a tuple*

$$policy = \langle pvalues, pa\_norms, pc\_norms, prules, impl \rangle$$

where:

- *pvalues is the set of* values *in the* Abstract Level *that are the origin of the policy,*

- *pa_norms is the set of* abstract norms *in the* Abstract Level *that are related to the policy,*

- *pc_norms is the set of* concrete norms *in the* Concrete Level *that complete the normative specification of the policy,*

- *prules is the set of* rules *in the* Rule Level *that describes the policy,*

- *impl is the implementation of the policy.*

5.7   Role Hierarchy

As mentioned in Section 5.2, the *objectives* in the organization's statutes can be represented in terms of goals (that we will call the *overall goal* of the society). As far as the organization has control over the actions of the agents acting within that organization, it will try to ensure that they perform actions that will lead to the overall goal of the society.

In our framework, the overall goals are the ones that help to define the role hierarchy. The process starts assigning the overall goals to the *root* of the role hierarchy, then subgoals are identified and goals are distributed among the different actors of the system. Therefore, goal division depends on the context.

Some of the criteria to be used in the goal division are the following:

1. One possible option is to divide goals into those that are the aims of the organization and those to be fulfilled by the agents entering the *e*-organization. This division defines two types of roles:

   - The *institutional roles*, that is, roles to be enacted by agents that represent and operate the *e*-organization, fulfilling all the goals of the organization. These corresponds to the *facilitation roles* defined by V.Dignum in [59].

   - The *external roles*, that is, roles to be enacted by agents representing external actors that enter into the organization.

   This means that agents performing an institutional role within the organization (such as *administrator*, *broker* or *auctioneer* in an auction house) will have a goal that is equal to some sub-goal of the goals of the organization. All the visiting agents performing external roles are expected to, at least, not obstruct the organization activity to obtain its main goal.

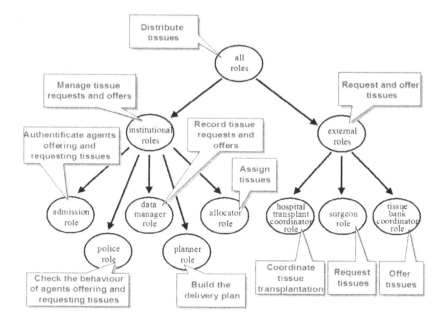

Figure 5.4: A subset of the role hierarchy with an example of goal distribution.

2. Another option is to refine a role by identifying sub-types of this role. This refinement defines a *is-a* relation between sub-roles and roles. For instance, if we have identified the *patient* role, we can refine it by sub-typing it into *donor* and *recipients*.[10]

3. We can also refine a role by decomposing it in sub-roles that, together, fulfill the goals of the given role. This refinement usually defines a *part-of* relation between sub-roles and roles. For instance, if we have identified *hospital* as a role with one or several goals to fulfill (e.g. search for donors), such role can be subdivided in sub-goals to be fulfilled by different staff members: *nurses*, *doctors* and *hospital transplant coordinator*.

Figure 5.4 shows as example part of the role hierarchy for the organ allocation problem[11]: we can see how there is an *overall goal* of the society placed on top of the hierarchy (*"Distribute Tissues"*), how this goal is splitted in the goal to be fulfilled by agents enacting institutional roles (*"Manage tissue requests and offers"*) and the ones enacting external roles (*"Request and offer tissues"*). Then those goals are divided iteratively in subgoals, and a role is identified to fulfill such goal.

---

[10]For the sake of simplicity, in the example we are assuming that *dead donnors* are also *patients*.

[11]A more complete version of this role hierarchy is presented in Section 6.5.2.

As we have seen in Section 3.3.1, another important relation among roles to be defined is the *power relation*. With this relation we empower some roles to be able to impose norms, actions or restrictions to other roles. In our framework we will use the power relation $\prec$ as a partial ordering among roles, which can be defined as a list of role pairs. In the case of the roles in figure 5.4, the list is as follows:

$$power\_rel = \quad \langle\langle police, admission\rangle , \langle police, data\_manager\rangle , \langle police, allocator\rangle ,$$
$$\langle police, planner\rangle , \langle allocator, planner\rangle , \langle police, external\_roles\rangle ,$$
$$\langle hospital\_transplant\_coordinator, surgeon\rangle\rangle$$

It is important to note here that, in the case of the $\langle police, external\_roles\rangle$ relation, we have used an intermediate node, as in this case the sub-roles of $external\_roles$ are connected through a *is-a* relation, therefore they inherit the power relation. This does not holds in the case of a *part-of* relation, as in such case the power relation might not apply to all the sub-roles but only to a subset.

Even though roles are highly coupled with goals, they are also tightly connected with the norms, rules and procedures defined in the Normative and Practical Systems of the *e*-organization. As the refinement process is made through the four levels, norms and rules are related to roles in the role hierarchy. For instance, let us have again a look to one of the norms defined in Section 5.2.2:

$$O_{ONT}(appropriate(distribution))$$

$$O_{hospital}(consent(donor) < done(transplant(hospital, donor, recipient, organ)))$$

The first norm is an abstract norm where the ONT role is defined (which corresponds to the *institutional roles* node in the hierarchy, as the ONT is the whole *e*-organization). The second norm is more concrete and refers to a *hospital* role (which corresponds to the *hospital transplant coordinator role* in the role hierarchy). For the sake of simplicity, in this chapter we used the ONT and *hospital* tags in the examples instead. The proper notation would be the following:

$$O_{inst}(appropriate(distribution))$$

$$O_{htc}(consent(donor) < done(transplant(htc, donor, recipient, organ)))$$

where $inst$ and htc are contractions of *institutional roles* and *hospital transplant coordinator role*, respectively.

The role hierarchy is, therefore, defined in parallel with the normative system. The abstract, concrete and rule levels reference to the roles can be used also to group norms and rules that are related with a given role. In Chapter 4 we defined $role_n$ as the set of rules that apply to an agent enacting a given role, but in fact there are also sets of abstract and concrete norms highly coupled with a given role. That means that the set $role_n$ could be defined at each level. All this is represented in figure 5.5, where roles are seen as another dimension in which the norms and rules are defined. Cutting that pyramid at different levels will result in definitions of roles in terms of norms and rules.

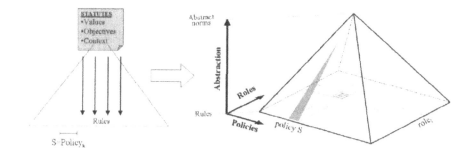

Figure 5.5: The relation among roles, policies and the abstraction levels.

As policies are another of the dimensions that are refined in the process, we can also identify the relation among roles and policies. Figure 5.5 shows how a role $role_i$ is related with a set of rules regarding a given policy S.

With all this in mind, we can define roles and the role hierarchy as follows:

Definition 5.9. *A Role Hierarchy of an e-Organization is a tuple*

$$role\_h = \langle roles, REL, org\_roles, external\_roles \rangle$$

where:

- $roles = \langle role_1, role_2, \ldots, role_n \rangle$ *is the set of roles defined in the role hierarchy,*

- $role_i = \langle goals, capabilities, ra\_norms, rc\_norms, r\_rules, capabilities,$ $protocols \rangle$ *is the complete definition of a single role,*

- *goals is the set of* goals *assigned to the role,*

- *capabilities is the set of* capabilities *agents enacting the role should have,*

- *ra_norms is the set of* abstract norms *in the* Abstract Level *that are related to the role,*

- *rc_norms is the set of* concrete norms *in the* Concrete Level *that are related to the role,*

- *r_rules is the set of* rules *in the* Rule Level *that constrain the behaviour of the agents enacting $role_i$,*

- *protocol is the set of protocols to be followed by the role,*

- $REL = \langle is\_a, part\_of, power \rangle$ *is a list of the binary relations between roles, (where is_a, part_of and power are pair lists $\langle role_i, role_j \rangle$),*

- *org_roles is the set of roles $role_i$ that compose the e-Organization,*

- *external_roles is the set of roles $role_i$ to be enacted by external agents that interact with the e-Organization.*

5.8 Ontologies

Apart from policies and roles, there is a third element that affects all four levels of the framework: ontologies. Ontologies are shared conceptualizations of terms and predicates in order to define a given domain. In our framework ontologies define the vocabulary to be used by all the agents in the *e*-organization. There are two main kinds of ontologies:

- *Domain ontologies*: these ontologies are focused to model the domain (e.g., organs and tissues) related with the *e*-organization activities.

- *Communication ontologies*: located in the procedure level, these ontologies define the predicates to allow agents to communicate one to the other (e.g., FIPA performatives).

In our framework *Domain ontologies* are defined in the abstract level and then extended in the following levels, as new norms and rules refer to new terms that are missing in the ontology. To give an example let us look again to the norms defined in Section 5.2.2:

$$O_{ONT}\left(appropriate(distribution)\right)$$

$$O_{hospital}\left(consent(donor) < transplant(hospital, donor, recipient, organ)\right)$$

The first norm refers to the term $distribution$, while the second one refers to $donor$, $recipient$ and $organ$. In both norms there is also a reference to some predicates: $appropriate$, $consent$ and $transplant$. All these terms and predicates are defined in the *Domain ontology*, and refined as the design process goes from abstract to concrete norms and rules.

5.9 Influence of the E-Organization's Context. Regulations

In Section 5.2, we saw how statutes make reference to a certain *context* where the organization performs its activities. In human societies the context of an organization includes regulations that are applied to the organization's internal and/or external behaviour. The main effect of those regulations for the designer is that they should be considered and included inside the designing process of the *e*-organization.

Figure 5.6 shows how the context in which the organization functions plays a role in our framework during the specification and implementation process. The abstractness of the norms and rules in those regulations decides at which level they affect the refinement process, as usually regulations define constraints in several levels, from the more abstract as *"do not discriminate because of race or sex"* to the rule level and, even, at the procedure level, by defining protocols to be followed as *"in situation A first do α, then..."*.

An example are the ONT statutes presented previously in Section 5.2. They clearly state that the ONT operates according to the spanish National Health system. The impact of being inside this parent organization is that, actually, the ONT should follow the regulations of the National Health system. Therefore, ONT *inherits* the norms and values of

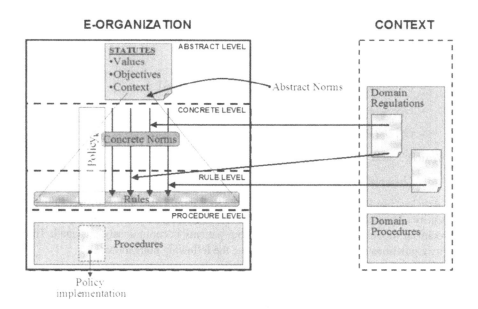

Figure 5.6: Influence of the virtual organization's context.

this system as well and they restrict the ways in which the objectives of the ONT can be reached.

Some of the *values* inherited from the National Health System are the following:

- *equity*: *"All citizens have the right to have Health services"* (Article 43 Spanish Constitution [44], Article 3.2 LGS[12] [114])

- *free service*; *"Donors and recipients should not pay or receive money"* (Article 2 Law 30/1979 [113], Article 8 Royal Decree 2070/1999 [180])

- *anonymity of donation*: *"It is forbidden to give or publish information allowing the identification of organ donors and recipients"* (Article 5 Royal Decree 2070/1999 [180])

- *no discrimination*: *"Patients should not be discriminated because of race, social status, sex, economical issues, ideology or political affiliation"* (Article 10.1 LGS [114])

The *Obligations* imposed to the ONT by the National Health System are stated in Royal Decree 2070/1999 [180]:

---

[12]Spanish National Health Law (*Ley General de Sanidad*).

- Article 19 states the following:

    - O_{ONT}(*Coordination of Organ and tissue allocation*)

    - O_{ONT}(*Organ and tissue transportation logistics*)

    - O_{ONT}(*Ensure quality and safety of organs and tissues*)

    - O_{ONT}(*Management of the recipient waiting lists*)

    - O_{ONT}(*Management of records about donors and recipients*)

- Article 2 adds the following obligation:

    - O_{ONT}(*Ensure equity in selection of recipients*)

Article 23 in RD 2070/1999 says that the ONT should also follow the regulations on security of personal records [118, 182], as ONT should manage waiting lists, donor and recipient records, where information about patients is continuously managed. The security regulations applied to the ONT impose the following obligations:

- O_{ONT}(*Ensure accuracy of all data about patients*)

- O_{ONT}(*Ensure security of data*)

- O_{ONT}(*Notify the APD about any new file with personal data before using it*)

- O_{ONT}(*Notify the APD about any modification in the structure of a file containing personal data before doing it*)

- O_{ONT}(*Notify the APD about the first transference of personal data to another organization*)

where APD is the Spanish Data Protection Agency (*Agencia de Protección de Datos*). An example of how the context affects the process has been already presented in previous sections. Figure 5.7 summarizes the example of the *appropriateness* of the distribution, and how it evolves through the different levels. In that example we already included the obligation

$$O_{ONT}(\textit{Ensure quality and safety of organs and tissues})$$

–which is one of the statements composing Article 19 in [180]– as part of the refinement process in the Concrete Level, splitting the *appropriate* predicate in two predicates: the *ensure_quality* and *ensure_compatibility* predicates.

The above shows that organizations hardly ever operate isolated and therefore frequent references are made to other regulations and organizations. ONT Statements also refer to *"ethical principles of equity"* as a criteria to be used in the organ allocation. Those ethics come from the society where the organization will operate, in this case the Spanish one (defined in the Spanish Constitution [44]).

The norms and values that are inherited from other organizations can be described at the top level of the *e*-organization explicitly. These norms have the highest priority and will *overwrite* any norms specified by the *e*-organization itself if they are contradictory. In our current work we are not considering those conflicting situations in the framework, but a quite simple solution to this scenario could be the use of a prioritized deontic logic, where default rules are used.

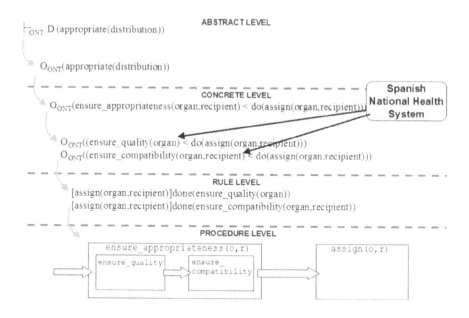

Figure 5.7: The influence of context in the refinement process.

5.10 Influence of the Background Knowledge

Implicitly in the process we described before there is also the influence of the background knowledge of the designer or designers of the *e*-organization, that acts as a repository where continuously solutions are fetched to solve a given issue.

Background knowledge is again affecting several levels. At the Abstract Level, this background knowledge is a source of ideas and generic structures to be followed. When defining an organization such as a new company, hardly no one creates the statutes and the organizational structure from scratch: usually are made by following/relying on some kind of common knowledge of how a company like the one we are willing to create should be structured, or either ideas are picked from other companies' statutes or organizational structures as example.

The same applies for the Concrete level, the Rule level and the Procedure level. An example is to design the search procedure to obtain a *correct distribution of organs*, or a *fair distribution*. As organs are a scarce resource, distribution is done one by one in a search process for a recipient. The background knowledge of the designer can tell which kind of search process could be implemented. One possible option is to simply translate the current ONT procedure, which is *appropriate* but not *efficient* in time. An alternative would be to use, for instance, the FIPA Call-For-Proposals (CFP) protocol, in order to do a more efficient distributed search for recipients.

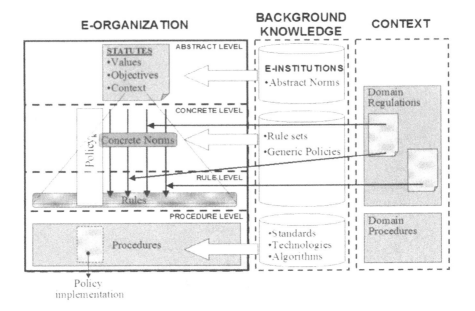

Figure 5.8: The Background Knowledge of the *e*-organization.

In our framework we propose to implement a background knowledge repository where solutions used in the modelling of previous *e*-organizations could be used again to create new ones. This idea is similar to the one in Object oriented approach, where a software company creates a Class library as repository to be used in the development of future software projects. This idea is depicted in figure 5.8, where the Background Knowledge is explicitly represented. A good example of elements in our framework that are quite reusable are the security policies: there is a quite similar structure from values to norms, to rules and the final implementation of by means of security mechanisms (such as encryption). These structures would be a kind of templates to be adapted for each case.

5.11   Creating Electronic Institutions

Let us now discuss again about the feasibility to build electronic institutions. As we explained in Section 3.4.3, using North's distinction between *institutions* and *organizations*, we cannot say that building a multi-agent system such as *CARREL* [48, 214] or the Fishmarket [72, 68] is the process to create an *e*-institution, but the one creating an *e*-organization.

However, in Section 5.10 we presented the idea of building a background knowledge repository. One of the aims of this idea is to provide the designer with templates that can be adapted to create new *e*-organizations.

In order to aid the designer to choose among the templates in the repository, some kind of guideline should be created. This is quite similar to the idea of V. Dignum (that we reviewed in Section 1.2.1) where, for each type of social structure, not only a template of the agent architecture and its related facilitation tasks is provided (*Matchmakers, Gatekeepers, Notaries, Controllers...*) but also a motivation to guide the choice among those structures based on the kind of problem to solve (*good exchange, collaboration, production*). The same could be applied to our background knowledge repository, creating abstract patterns and templates that are structured in a way that helps the designer in the design process. It is the process to build those abstract patterns and templates the one that is closer to the idea of institutions as abstract entities.

With this idea in mind we can define an *Electronic Institution* as follows:

Definition 5.10. *An Electronic Institution (e-institution) is a set of templates that can be adapted, parameterized or instantiated to build an e-organization.*

5.12 Summary

In this chapter we extend the HARMON*IA* framework introduced in Chapter 4 by incorporating the *e*-organization perspective. We have identified the different levels of abstraction that compose the normative framework:

- An *Abstract Level*, where the *values*, *objectives* and *context* of the *e*-organization are defined (by its statutes) and then translated in terms of *abstract norms*;

- A *Concrete Level*, where *abstract norms* are refined into more *concrete norms* in order to reduce the abstractness of the normative system, by fixing the interpretation of some predicates and terms in the context of the *e*-organization;

- A *Rule Level*, where concrete norms are translated into rules to be computed by agents. This translation is needed because, as we have seen in Section 3.2.1, norms have no operational semantics (i.e., they only define what *ought to be*, but not *how to be done*);

- A *Procedure Level*, where the final mechanisms to implement the rules are created. We propose to use a) *rule interpreters* to create agents able to reason about the rules, and b) to translate the rules into protocols. With this dual approach we allow incoming agents (i.e., agents entering into the *e*-organization) to either follow the protocol, the rules, or both, depending on their capabilities. We introduce the *Flexible Normative Agents*, which are able to choose, in each moment, to either follow the protocol or to use the rules in order to cope with those situations the protocol did not foresee.

The abstract and concrete levels compose the *Normative System* of the *e*-organization, while the rule and procedure levels compose the *Practical System*. This division of

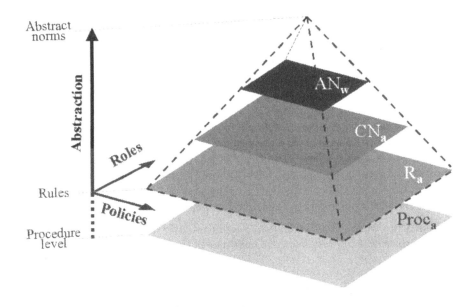

Figure 5.9: The relation among the organizational view and the agent view.

levels remarks the distinction between the *normative dimension* of the *e*-organization (the one that states what is acceptable or unacceptable by means of *obligations, permissions* and *rights*) and the practical, *descriptive dimension*, (the one defining the *praxis*, that is, how agents should behave in order to meet the norms).

Then we describe the refinement process to build an agent-mediated normative electronic organization, from the abstractness of the organization's statutes to the norms to the final implementation of those norms in rules and procedures.

These four levels of abstraction in the normative framework correspond to the different levels of normative abstraction that we identified in Chapter 4 from the agent's point of view. Figure 5.9 shows the relation among both views, as the normative and descriptive systems correspond to the sets AN_w, CN_a, R_a and $Proc_a$ defined in Section 4.2.

Apart of the aforementioned horizontal levels, the framework also identifies some vertical components that are defined through different levels:

- The *Policies*, which can be traced from the *values* in the abstract level to the defined *rules* and *violations* in the rule level and then to their translation into *procedures* (the so-called *implementation of the policy*).

- The *Role Hierarchy*, which is defined in the abstract level as the *objectives* of the *e*-organization, and then is refined in the next levels by a careful process of *goal distribution* that is used to define the *roles* in the *e*-organization. Role definition is done in parallel with the refinement of norms, by defining the roles that are affected by a given norm (this is done in order to solve the *role abstraction* of norms).

- The *Ontologies*, which define the concepts needed in the *e*-organization. In our framework we identify two kinds of ontologies:

  - The *Domain Ontology*, which defines the *terms, predicates* and *actions* that appear in the abstract, concrete and rule level. The ontology is defined in parallel to the refinement process of the normative system, in order to solve the abstraction of norms coming from vague terms or actions.

  - The *Communication Ontology*, which defines the *performatives* to be used by the process agents in the procedure level.

- The surrounding *Context* of the *e*-organization, which influences in the refinement process of the normative framework at different levels of abstraction, from abstract *values* to complete *procedures* to be followed.

- The *Background Knowledge*, a repository where the designer can fetch solutions, patterns and procedures to be applied to the refinement process. Our idea is to build such a repository to store previous solutions used i order to be re-sued again in the development of further *e*-organizations, such as the elements related to security issues.

During the description of the aforementioned levels and components, the terminology proposed in Chapter 4 is extended to include concepts such as *statutes, values, role hierarchy, policy,* or to define the difference between *e-organizations* and *e-institutions*: in our framework, an *e*-institution is a set of templates that can be adapted, parameterized or instantiated to build an *e*-organization. Such templates are located in the repository that composes the *background knowledge*.

In the next chapter we will extensively show the use of HARMON*IA* to model the organ and tissue allocation problem, and we will demonstrate its usefulness to determine if a given multi-agent system follows a complex normative framework.

Chapter 6

Applying HARMON*IA* to the Organ and Tissue Allocation Problem

In previous chapters we have presented HARMON*IA* , a framework to model electronic organizations, from the abstract level where norms usually are defined to the final protocols and procedures that implement the norms. In Chapter 5 we also presented a small example about organ and tissue allocation to illustrate some of the concepts.

In this chapter we will extend the organ and tissue example in order to describe, more extensively, the steps that compose the refining process. The aim of this example is twofold: i) to show how the framework can be applied to a real domain, and ii) to analyse the *CARREL* implementation presented in Chapter 2, check if it follows all the regulations and, even, find elements which should be added to the system.

The example will be extensive but not complete. We will mainly focus on the normative dimension of the process, as it is the most innovative part of our framework. We will also show how the role hierarchy is defined and how the main goals are distributed among the roles.

Abbreviations

During this chapter we will use the following abbreviations:

- *ONT*: the Spanish National Transplant Organization,

- *SpNHS*: the Spanish National Health System,

- *APD*: the Spanish Data Protection Agency,

- *SpLaw*: the Spanish Law,

- *EULaw*: the European Union Law,

- *Const*: the Spanish Constitution [44],

- *LGS*: the Spanish National Health Law [114],

- *L.30.79*: Law 30/1979 [113] (organ transplantation).

- *RD.2070.99*: the Royal Decree 2070/1999 [180] (organ and tissue allocation),

- *LOPD*: the Spanish Organic Law about Data Protection [118],

- *RD.994.99*: the Royal Decree 944/1999 [182] (security measures),

- *R.97.5*: European Recommendation R(97)5 [183] (security measures), and

- *CRL*: for the *CARREL* context.

Notation

In Section 4.2.2 we defined context as boundaries that define 1) a shared vocabulary and b)a normative framework to be followed by a certain group of individuals (definition 4.7). Within this view, a human organization and an *e*-organization are both contexts. In Section 5.9 we referred to the influence of the surrounding context of the *e*-organization. Both views of contexts will be used in this chapter. In order to avoid confusion, we will refer to both views as follows:

- "the e-organization context": in this case we refer to the context that defines the vocabulary and the normative framework of the e-organization. In figure 6.1:

    – "the e-org$_x$ context" would refer to the context tagged as "e-org$_x$".

    – "the $CARREL$ context" would refer to the context tagged as $CARREL$.

- "the context of the e-organization": in this case we refer to those super-contexts that affect the e-organization context. In figure 6.1:

    – "the context of e-org$_x$" would refer to the set of super-contexts of "e-org$_x$", that is, from "C_a" to "C_n".

    – "the context of $CARREL$" would refer to the following super-contexts: "ONT", "SpNHS", "SpLaw" and "EULaw". Similarly, "the context of ONT" would refer to the "SpNHS", "SpLaw" and "EULaw" contexts.

6.1   The Abstract Level in *CARREL*

The *CARREL* system is a multi-agent system designed to help a transplant organization to allocate organs and tissues (in our case, the Spanish *National Transplant Organization*, that we will refer from now on as ONT). Therefore, such a system should include the ONT's statutes as part of its specification, as it should share the same values, objectives and context.

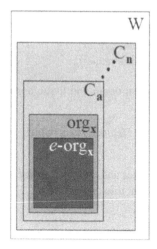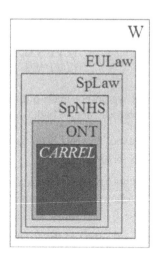

Figure 6.1: Graphical representation of the nested contexts of a) (left) a generic *e*-organization, and b) (right) the *CARREL* *e*-organization .

6.1.1 Defining ONT Statutes

The first step is to describe the *statutes* of the *e*-organization, identifying the *values, objectives and context*. As described in Section 5.2, the ONT statutes are as follows:

> *The principal objective of the ONT is therefore the promotion of donation and the consequent* **increase of organs** *available for transplantation, from which all its other functions result. The ONT acts as a service agency for the whole* **National Health System,** *works for the continuing increase in the availability of* **organs and tissues for transplantation** *and guarantees the most* **appropriate and correct distribution,** *in accordance with the degree of technical knowledge and ethical principles of* **equity** *which should prevail in the transplant activity.*

From this statement we can define that:

- The values of ONT (expressed as desires) are the following:

$$ONT:D1 \quad \vdash_{ONT} D(appropriate(distribution))$$
$$ONT:D2 \quad \vdash_{ONT} D(correct(distribution))$$
$$ONT:D3 \quad \vdash_{ONT} D(equal(distribution))$$

- The objectives of ONT (expressed as goals) are the following:

$$ONT{:}O1 \quad \vdash_{ONT} G(increase(donations))$$
$$ONT{:}O2 \quad \vdash_{ONT} G(allocate(organs))$$
$$ONT{:}O3 \quad \vdash_{ONT} G(allocate(tissues))$$

Note that we do not make a difference between organ and tissue donations, as a single donor can donate both.

- The context of the ONT is the *Spanish National Health System* (which we will refer from now on as SpNHS)

Note that we have given an identifier to the values and the goals, in order to be able to track in the following the origin of any value, norm, rule or procedure.

Influence of the Context of the E-Organization

As explained in Section 5.9, the context of the *e*-organization may affect its specification at different levels. In this case, the context of ONT is the SpNHS, SpLaw and EULaw. During the process we will continuously check if these super-contexts may define a restriction that affects the ongoing specification.

At the level of values, the SpNHS defines some values that are inherited by the ONT:

$$RD.2070.99{:}5 \quad \vdash_{SpNHS} D(anonymity(patient))$$
$$LGS{:}3.2 \quad \vdash_{SpNHS} D(equity(patient, service))$$

The RD.2070.99:5 value[1] states that anonymity of all patients should be kept. This value is not covered by any ONT value. Therefore, new values should be added to the list of ONT values:

$$ONT{:}D4.1 \quad \vdash_{ONT} D(anonymity(donor)) \Rightarrow_{ONT} RD.2070.99{:}5$$
$$ONT{:}D4.2 \quad \vdash_{ONT} D(anonymity(recipient)) \Rightarrow_{ONT} RD.2070.99{:}5$$

Note here the use of the *counts-as* operator at the level of values. It is used to say that a value (such as ONT:D4.1) *counts-as* (in the ONT context) the value expressed in RD.2070.99:5. (i.e., ONT:D4.1 fixes the interpretation of RD.2070.99:5 in the ONT context). It is also important to note here that $ONT : D4.1 \Rightarrow_{ONT} ONT : D4$ and $ONT : D4.2 \Rightarrow_{ONT} ONT : D4$. In order to be concise, we will not add these relations, that can be inferred by the naming convention we have chosen.

As well, LGS:3.2 value (i.e., value coming from Article 3.2 in LGS) states that equity should prevail in the access of all patients to the health services. This one is already covered by ONT:D3. But it is important to note here that LGS:3.2 value has been refined

---

[1]Notation: for values, norms and rules coming from an article in a given law or regulation, we will use the notation *law : article*. For instance, RD.2070.99:5 means *Article 5 in RD.2070.99*, while LGS:3.2 means *Article 3.2 in LGS*.

by Article 10.1 in LGS as follows:

$LGS:10.1a$ $\vdash_{SpNHS} D(no\_discrimination(patient, race)) \Rightarrow_{SpNHS} LGS:3.2$
$LGS:10.1b$ $\vdash_{SpNHS} D(no\_discrimination(patient, sex)) \Rightarrow_{SpNHS} LGS:3.2$
$LGS:10.1c$ $\vdash_{SpNHS} D(no\_discrimination(patient, ideology)) \Rightarrow_{SpNHS} LGS:3.2$
$LGS:10.1d$ $\vdash_{SpNHS} D(no\_discrimination(patient, social\_status))$
$\Rightarrow_{SpNHS} LGS:3.2$

expressing that no patient should be discriminated because of race, sex, ideology or social status. Note here the use of the $\Rightarrow$ operator to say that the interpretation of LGS:3.2 done by LGS:10.1x is done in the SpNHS context. Further laws refined LGS:10.1d stating that organ and tissue transplantation should be a free service. We can express that as follows:

$RD.2070.99:8$ $\vdash_{SpNHS} D(no\_money(patients, transplant)) \Rightarrow_{SpNHS} LGS:10.1d$

With all that in mind, ONT:D3 should be refined as follows:

$ONT:D3.1$ $\vdash_{ONT} D(no\_discrimination(recipient, race)) \Rightarrow_{ONT} LGS:10.1a$
$ONT:D3.2$ $\vdash_{ONT} D(no\_discrimination(recipient, sex)) \Rightarrow_{ONT} LGS:10.1b$
$ONT:D3.3$ $\vdash_{ONT} D(no\_discrimination(recipient, ideology)) \Rightarrow_{ONT} LGS:10.1c$
$ONT:D3.4$ $\vdash_{ONT} D(no\_money(donor, piece)) \Rightarrow_{ONT} RD.2070.99:8$
$ONT:D3.5$ $\vdash_{ONT} D(no\_money(recipient, piece)) \Rightarrow_{ONT} RD.2070.99:8$

Identifying the Policies in *CARREL*

At the level of values we can already identify the roots of the different policies of the *CARREL* system:

- the *organ allocation* and *tissue allocation policies* have its roots on values ONT:D1, ONT:D2, ONT:D3.1, ONT:D3.2, ONT:D3.3, ONT:D3.4 and ONT:D3.5;

- the *security policy* has its roots on values ONT:D2, ONT:D4.1 and ONT:D4.2.

In following sections we will refine each of those policies, as we define their norms, rules and procedures.

6.1.2 Defining the Abstract Norms

The next step is the definition of the Abstract Norms from the values previously identified.

$ONT:A1$ $O_{ONT}(appropriate(distribution)) \mapsto_{ONT} ONT:D1$
$ONT:A2$ $O_{ONT}(correct(distribution)) \mapsto_{ONT} ONT:D2$
$ONT:A3.1$ $O_{ONT}(no\_discrimination(recipient, race)) \mapsto_{ONT} ONT:D3.1$
$ONT:A3.2$ $O_{ONT}(no\_discrimination(recipient, sex)) \mapsto_{ONT} ONT:D3.2$
$ONT:A3.3$ $O_{ONT}(no\_discrimination(recipient, ideology)) \mapsto_{ONT} ONT:D3.3$
$ONT:A3.4$ $O_{ONT}(no\_money(donor, piece)) \mapsto_{ONT} ONT:D3.4$
$ONT:A3.5$ $O_{ONT}(no\_money(recipient, piece)) \mapsto_{ONT} ONT:D3.5$
$ONT:A4.1$ $O_{ONT}(anonymity(donor)) \mapsto_{ONT} ONT:D4.1$
$ONT:A4.2$ $O_{ONT}(anonymity(recipient)) \mapsto_{ONT} ONT:D4.2$

Note that in this case we have not used the *counts-as* operator ($\Rightarrow$), as, in this case, we have translated from the language of values to the language of abstract norms. Therefore, we use a translation operator $\mapsto$ which allows us to keep track of the transition.

Influence of the Context of the E-Organization

Again the surrounding context may add new abstract norms in this level. In this case, as mentioned in Section 5.9, Article 19 in RD.2070.99 defines the following obligations:

$$RD.2070.99:19a \quad O_{ONT}(coordinate(distribution))$$
$$RD.2070.99:19b \quad O_{ONT}(organize\_logistics(distribution))$$
$$RD.2070.99:19c \quad O_{ONT}(ensure\_quality(piece) \wedge ensure\_safety(piece))$$
$$RD.2070.99:19d \quad O_{ONT}(management(recipient\_waiting\_lists))$$
$$RD.2070.99:19e \quad O_{ONT}(record(donors) \wedge record(recipients))$$
$$RD.2070.99:19f \quad O_{ONT}(ensure\_equity(distribution))$$

In the previous list we can identify 2 groups of norms:

- new norms: norms that are not covered by the ones previously identified. In this case those norms are RD.2070.99:19b, RD.2070.99:19d and RD.2070.99:19e. Therefore, they are automatically introduced in the set of abstract norms of the ONT:

$$ONT:A5 \quad O_{ONT}(organize\_logistics(distribution)) \Rightarrow_{ONT} RD.2070.99:19b$$
$$ONT:A6 \quad O_{ONT}(management(recipient\_waiting\_lists))$$
$$\Rightarrow_{ONT} RD.2070.99:19d$$
$$ONT:A7.1 \quad O_{ONT}(record(donors)) \Rightarrow_{ONT} RD.2070.99:19e$$
$$ONT:A7.2 \quad O_{ONT}(record(recipients)) \Rightarrow_{ONT} RD.2070.99:19e$$

- related norms: norms that are related with some of the ones previously identified. In this case:

  - RD.2070.99:19a states that ONT should coordinate the distribution process for transplants. This fact is already expressed by ONT:A1 and ONT:A2, which state the obligation of ONT to coordinate the distribution process ensuring appropriateness and correctness. This event of finding norms already covered by the specification should not be seen as a negative fact making the designer to revisit norms already identified. In these cases we are, in fact, discovering the norms which were the origin of the values and norms defined by the organization. In this case we have found the norm which is the origin of some of the ONT norms we already identified. Therefore, we will add this relationship through the *counts-as* operator:

$$ONT:A1 \quad O_{ONT}(appropriate(distribution)) \Rightarrow_{ONT} RD.2070.99:19a$$
$$ONT:A2 \quad O_{ONT}(correct(distribution)) \Rightarrow_{ONT} RD.2070.99:19a$$

- RD.2070.99:19f refers to equity in the distribution. This norm is already covered by ONT:A3, so we are in a case similar to the previous one. The resulting modification in the norm specification is the following:

$$ONT:A3 \quad O_{ONT}(equity(distribution)) \Rightarrow_{ONT} RD.2070.99:19f$$

It is important to note here that we have only placed the relationship between ONT:A3 and RD.2070.99:19f. It is not necessary to place such relation with the norms derived from ONT:A3, as the *counts-as* operator is transitive:

$$ONT:A3.1 \Rightarrow_{ONT} ONT:A3 \Rightarrow_{ONT} RD.2070.99:19f$$

- RD.2070.99:19b states that the quality and safety of pieces (organs and tissues) should be ensured. In this case this norm is related with the appropriateness of pieces, so this norm is refining ONT:A1. The result of introducing this norm is the following:

$$ONT:A1.1 \quad O_{ONT}(ensure\_quality(piece)) \Rightarrow_{ONT} RD.2070.99:19b$$
$$ONT:A1.2 \quad O_{ONT}(ensure\_safety(piece)) \Rightarrow_{ONT} RD.2070.99:19b$$

It is important to note that $ONT:A1.1 \Rightarrow_{ONT} ONT:A1$ and $ONT:A1.2 \Rightarrow_{ONT} ONT:A1$. In order to be concise, we will not add these relations, that can be inferred by the naming convention we have chosen.

6.1.3 Summary: the Abstract Level in *CARREL*

With all this the abstract level of *CARREL* can be described as follows (see *definition 5.4*):

$abslevel = \langle statutes, a\_norms, ontology_{al} \rangle$, where:

- $statutes = \langle values, objectives, context \rangle$

  - $values = \langle value_1, value_2, value_{3.1}, value_{3.2}, value_{3.3}, value_{3.4}, value_{3.5}, value_{4.1}, value_{4.2} \rangle$
    * $value_1 = \langle ONT:D1, \langle ONT:A1.1, ONT:A1.2 \rangle \rangle$
    * $value_2 = \langle ONT:D2, ONT:A2 \rangle$
    * $value_{3.1} = \langle ONT:D3.1, ONT:A3.1 \rangle$
    * $value_{3.2} = \langle ONT:D3.2, ONT:A3.2 \rangle$
    * $value_{3.3} = \langle ONT:D3.3, ONT:A3.3 \rangle$
    * $value_{3.4} = \langle ONT:D3.4, ONT:A3.4 \rangle$
    * $value_{3.5} = \langle ONT:D3.5, ONT:A3.5 \rangle$
    * $value_{4.1} = \langle ONT:D4.1, ONT:A4.1 \rangle$
    * $value_{4.2} = \langle ONT:D4.2, ONT:A4.2 \rangle$

    – $objectives = \langle ONT\!:\!O1, ONT\!:\!O2, ONT\!:\!O3 \rangle$

    – $context = \langle SpNHS \rangle$

- $a\_norms = \langle ONT\!:\!A1.1, ONT\!:\!A1.2, ONT\!:\!A2, ONT\!:\!A3.1,$
  $ONT\!:\!A3.2, ONT\!:\!A3.3, ONT\!:\!A3.4, ONT\!:\!A3.5, ONT\!:\!A4.1, ONT\!:\!A4.2$
  $ONT\!:\!A5, ONT\!:\!A6, ONT\!:\!A7.1, ONT\!:\!A7.2 \rangle$

- $ontology_{al}$ is the ontology that defines all the terms and predicates that have appeared so far.

6.2   Role Hierarchy and Goal Distribution

A task to be performed in parallel during all the refinement process is the definition of the role hierarchy. As mentioned in Section 6.1.1, the role hierarchy design is mainly guided by the goal distribution that comes from the objectives identified in the statutes.

In Section 6.1.1, the identified goals of the ONT were the following:

$$ONT\!:\!O1 \quad \vdash_{ONT} \ G(increase(donations))$$
$$ONT\!:\!O2 \quad \vdash_{ONT} \ G(allocate(organs))$$
$$ONT\!:\!O3 \quad \vdash_{ONT} \ G(allocate(tissues))$$

The first goal (ONT:O1) is to *increase donations*. This objective is out of the scope of any software, as this goal involves:

- the patients' will (expressed by themselves or through their families) to donate organs,

- the work of the people in hospitals, looking for donors, and

- the work of the staff at the ONT, promoting donation through several ways, including the media.

The result of this analysis gives the role hierarchy depicted in figure 6.2, where we distinguish among the *patients* (which give their consent), and the *health organizations* (which seek donations). In the case of patients we distinguish also between *donors* (the patients which give the consent) and *recipients* (which have no contribution to the *give consent* goal but are introduced for completeness). In the case of the *health organizations* we also distinguish among the hospitals (which should identify possible donors) and the ONT (the one which promotes donation). Finally, we refine the *ONT* role, stating that it is composed by ONT's *board* –the one who cares about promotion of donation– and the *CARREL* system (which has no contribution to that goal, as promotion is outside the scope of the system).

The second goal to analyse is *distribute tissues*. The process starts again in the root node, and then it distributes the goal among the roles already defined, and adds new roles when they are needed. We can see the result of this process in figure 6.3. At the root node we find that *patients* have no contribution to the *distribute tissues* goal, so in this

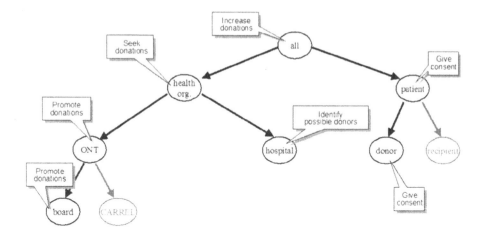

Figure 6.2: Role hierarchy resulting from the distribution of the *increase donations* goal.

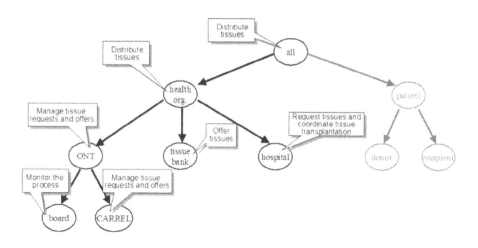

Figure 6.3: Role hierarchy resulting from the distribution of the *distribute tissues* goal.

case the goal passes unchanged to the *health organization* role. Then we distribute the goal between the *ONT* (which manages the requests and offers for tissues), the *hospitals* (which may request for tissues and also have to coordinate the transplantation process inside the hospital) and the *tissue banks* (which are the ones that offer tissues to *ONT*). The last one is a missing role, so it is added to the hierarchy. To finalize the distribution, the *manage tissue requests and offers* goal is mainly given to *CARREL*, but ONT's *board* has also a contribution as it has to monitor the process.

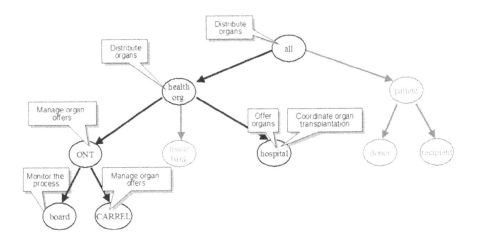

Figure 6.4: Role hierarchy resulting from the distribution of the *distribute organs* goal.

The last goal to analyse is *distribute organs*. As in the previous case, the process uses the role hierarchy already defined and starts in its root node. The process (depicted in figure 6.4) is almost identical to the one for tissue distribution, with the difference that tissue banks have no contribution to this goal, as organs go directly from one hospital where the donor is) to the same or another hospital (where the recipient is).

In the hierarchy obtained from the distribution of the three goals of the ONT, we can identify two different relations among roles:

- *part-of relations*: the ones from the *ONT* to the *board* and *CARREL* roles, as both together compose the ONT.

- *is-a relations*: the rest of relations.

As described in Section 5.7, we should also define the *power relations* between roles, in order to describe which roles are empowered to impose norms, actions or restrictions to other roles. In this case is clear that ONT is empowered (by Spanish law) over hospitals and tissue banks. Then we should add this relations. The next step is to analyse how the power relation affects the *board* and CARREL roles. As in this case there is a *part-of*[2] relation, we have no automatic inheritance of the power relations, so we have to determine them one by one. In this case both the ONT board and the CARREL system receive the power relations of the ONT role, as both have to deal with hospitals and tissue banks although at different levels. The resulting power relations, depicted in figure 6.5, are the following:

---

[2]In the case of *is-a* relations, the sub-roles inherit all the power relations of the super-role. In the case of *part-of* relations, the power relations of the super-role are distributed among the sub-roles, as the power relation might not apply to all the subroles but only a subset.

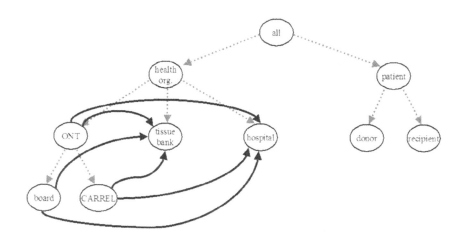

Figure 6.5: Power relations at the role hierarchy.

$$power\_rel = \quad \langle\langle ONT, hospital\rangle, \langle ONT, tissue\_bank\rangle, \langle board, hospital\rangle,$$
$$\langle board, tissue\_bank\rangle, \langle CARREL, hospital\rangle,$$
$$\langle CARREL, tissue\_bank\rangle\rangle$$

6.3 The Concrete Level in *CARREL*

To start the definition of the Concrete Level, the initial set of Concrete Norms should be identified. The set comes from the Abstract Norms identified in the Abstract Level. In this case, the resulting set of concrete norms is the following:

$ONT:C1.1 \quad O_{ONT}(ensure\_quality(piece)) \equiv ONT:A1.1$
$ONT:C1.2 \quad O_{ONT}(ensure\_safety(piece)) \equiv ONT:A1.2$
$ONT:C2 \quad O_{ONT}(correct(distribution)) \equiv ONT:A2$
$ONT:C3.1 \quad O_{ONT}(no\_discrimination(recipient, race)) \equiv ONT:A3.1$
$ONT:C3.2 \quad O_{ONT}(no\_discrimination(recipient, sex)) \equiv ONT:A3.2$
$ONT:C3.3 \quad O_{ONT}(no\_discrimination(recipient, ideology)) \equiv ONT:A3.3$
$ONT:C3.4 \quad O_{ONT}(no\_money(donor,piece)) \equiv ONT:A3.4$
$ONT:C3.5 \quad O_{ONT}(no\_money(recipient, piece)) \equiv ONT:A3.5$
$ONT:C4.1 \quad O_{ONT}(anonymity(donor)) \equiv ONT:A4.1$
$ONT:C4.2 \quad O_{ONT}(anonymity(recipient)) \equiv ONT:A4.2$
$ONT:C5 \quad O_{ONT}(organize\_logistics(distribution)) \equiv ONT:A5$
$ONT:C6 \quad O_{ONT}(management(recipient\_waiting\_lists)) \equiv ONT:A6$
$ONT:C7.1 \quad O_{ONT}(record(donors)) \equiv ONT:A7.1$
$ONT:C7.2 \quad O_{ONT}(record(recipients)) \equiv ONT:A7.2$

It is important to note here that we have used the $\equiv$ relation instead of the $\Rightarrow$ (counts-as) or $\rightarrow$ (translation). This is because, as mentioned in Section 5.2.2, we use the same language for abstract and concrete norms (i.e., there is no translation), so we have not made a new interpretation of the norms in the abstract level (i.e., there is no change from the context where the previous norms were defined to a new one). But a translation is needed at this point in those cases where the language for Abstract Norms ($ANorms$) and the one for Concrete Norms ($CNorms$) are different.

Once the initial set of Concrete Norms is defined, the refining process begins. In this process we will reduce the abstraction of the norms (i.e., abstract actions, vague terms, temporal abstractness, agent and role abstraction or unverifiable actions, as seen in 5.3.1).

In order to make the example easier to understand, in the following sections we will divide the norms in terms of the policies they belong to.

6.3.1 The Organ and Tissue Allocation Policies

At this level, the norms defining the organ and tissue policies are the following:

$$ONT\!:\!C1.1 \quad O_{ONT}(ensure\_quality(piece))$$
$$ONT\!:\!C1.2 \quad O_{ONT}(ensure\_safety(piece))$$
$$ONT\!:\!C2 \quad O_{ONT}(correct(distribution))$$
$$ONT\!:\!C3.1 \quad O_{ONT}(no\_discrimination(recipient, race))$$
$$ONT\!:\!C3.2 \quad O_{ONT}(no\_discrimination(recipient, sex))$$
$$ONT\!:\!C3.3 \quad O_{ONT}(no\_discrimination(recipient, ideology))$$
$$ONT\!:\!C3.4 \quad O_{ONT}(no\_money(donor, piece))$$
$$ONT\!:\!C3.5 \quad O_{ONT}(no\_money(recipient, piece))$$
$$ONT\!:\!C5 \quad O_{ONT}(organize\_logistics(distribution))$$

These norms are the ones stating ONT's obligations such as ensuring quality and safety of the pieces, avoiding discrimination and monetary transactions, or its role in the organization of the logistics to transfer organs and tissues from one place to the other.

As mentioned in Section 5.3.1, the norms coming from the Abstract Level use to be far too abstract to be useful. So these norms should be refined to reduce their abstraction.

$$ONT\!:\!C1.1.1 \quad O_{ONT}(ensure\_quality(piece) < do(assign(piece, recipient)))$$
$$ONT\!:\!C1.1.1.1 \quad O_{origin\_org}(ensure\_quality(piece))$$
$$ONT\!:\!C1.1.1.1.1 \quad O_{hospital}(ensure\_quality(organ))$$
$$ONT\!:\!C1.1.1.1.2 \quad O_{tissue\_bank}(ensure\_quality(tissue))$$
$$ONT\!:\!C1.1.2 \quad O_{CARREL}(get\_quality(piece, origin\_org)$$
$$< do(assign(piece, recipient)))$$
$$ONT\!:\!C1.1.2.1 \quad O_{CARREL}(get\_quality(organ, hospital)$$
$$< do(assign(organ, recipient)))$$

$ONT:C1.1.1.2.2 \quad O_{CARREL}(get\_quality(tissue, tissue\_bank)$
$< do(assign(tissue, recipient)))$

$ONT:C1.2.1 \quad O_{ONT}(ensure\_compatibility(piece, recipient)$
$< do(assign(piece, recipient)))$

$ONT:C1.2.1.1 \quad O_{CARREL}(ensure\_compatibility(piece, recipient)$
$< do(assign(piece, recipient)))$

$ONT:C1.2.1.1.1 \quad O_{CARREL}(ensure\_compatibility(organ, recipient)$
$< do(assign(organ, recipient)))$

$ONT:C1.2.1.1.2 \quad O_{CARREL}(ensure\_compatibility(tissue, recipient)$
$< do(assign(tissue, recipient)))$

This first set of norms refines the ones related to the quality and the compatibility of organs and tissues. In the case of norm ONT:C1.1.1, first we have introduced a time constraint (quality assessment should be done before the piece is assigned). Then a separation of duties is made among roles. For the organizations that are originators of the piece, as they do not perform the assignation, they only have the obligation to ensure quality of pieces, while for *CARREL* the quality assessment is reduced to get such information from the organization (hospital or tissue bank) which is the origin of the piece. The information about the quality of piece should be obtained before the assignation is made, introducing here a restriction to reduce temporal abstractness. A similar process is made for obligation ONT:C1.2. In this case the obligation to ensure compatibility only lays on *CARREL*, as it is the one performing the assignation.

$ONT:C2 \quad O_{ONT}(correct(distribution))$
$ONT:C2.1 \quad O_{ONT}(efficient(distribution))$
$ONT:C2.1.1 \quad O_{ONT}(time\_efficient(distribution))$
$ONT:C2.1.1.1 \quad O_{CARREL}(seek\_time\_efficiency(distribution))$

These norms refine the norm about correctness of the distribution. In this case, we first interpret *correct* distribution as one that is efficient, and then we specify which kind of efficiency we are interested in. In the case of organs and tissues, *time* is one of the major constraints, as the quality of organs and some tissues is inverse to the time they are preserved outside a human body.

$ONT:C3.1.1 \quad O_{CARREL}(no\_discrimination(recipient, race))$
$ONT:C3.2.1.1 \quad F_{CARREL}(assign(piece, recipient) \wedge use(recipient, race))$
$ONT:C3.2.1.1.1 \quad F_{CARREL}(assign(organ, recipient) \wedge use(recipient, race))$
$ONT:C3.2.1.1.2 \quad F_{CARREL}(assign(tissue, recipient) \wedge use(recipient, race))$
$ONT:C3.2.1 \quad O_{CARREL}(no\_discrimination(recipient, sex))$
$ONT:C3.2.1.2 \quad F_{CARREL}(assign(piece, recipient) \wedge use(recipient, sex))$
$ONT:C3.2.1.2.1 \quad F_{CARREL}(assign(organ, recipient) \wedge use(recipient, sex))$
$ONT:C3.2.1.2.2 \quad F_{CARREL}(assign(tissue, recipient) \wedge use(recipient, sex))$
$ONT:C3.3.1 \quad O_{board}(no\_discrimination(recipient, ideology))$

This set refines norms about non-discrimination of recipients. The refinement first identified which are the roles related to each obligation. In the case of ideology we considered that this information will not be in the data the *CARREL* system receives about patients, so we derived the obligation to ONT's board. In the cases of race or sex, those fields might be included in a patient's record. Therefore, we should forbid *CARREL* to use such information in the assignation.

$$ONT\!:\!C3.4.1 \quad O_{board}(no\_money(donor, piece))$$
$$ONT\!:\!C3.4.1.1 \quad O_{board}(no\_money(donor, organ))$$
$$ONT\!:\!C3.4.1.2 \quad O_{board}(no\_money(donor, tissue))$$
$$ONT\!:\!C3.4.2 \quad F_{donor}(receive\_money(donor, piece))$$
$$ONT\!:\!C3.4.2.1 \quad F_{donor}(receive\_money(donor, organ))$$
$$ONT\!:\!C3.4.2.2 \quad F_{donor}(receive\_money(donor, tissue))$$
$$ONT\!:\!C3.5.1 \quad O_{board}(no\_money(recipient, piece))$$
$$ONT\!:\!C3.5.1.1 \quad O_{board}(no\_money(recipient, organ))$$
$$ONT\!:\!C3.5.1.2 \quad O_{board}(no\_money(recipient, tissue))$$
$$ONT\!:\!C3.5.2 \quad F_{all}(receive\_money(recipient, piece))$$
$$ONT\!:\!C3.5.2.1 \quad F_{all}(receive\_money(recipient, organ))$$
$$ONT\!:\!C3.5.2.2 \quad F_{all}(receive\_money(recipient, tissue))$$

These norms are related with the intention of avoiding any payment done to donors or by recipients. In this case, as we do not pretend *CARREL* to be a system able to check the patients' bank accounts, enforcement of these norms is responsibility of the ONT's board.

$$ONT\!:\!C5.1.1 \quad O_{ONT}(organize\_logistics(piece, origin\_org, dest\_org))$$
$$ONT\!:\!C5.1.1.1 \quad O_{CARREL}(plan\_delivery(piece, origin\_org, dest\_org))$$
$$ONT\!:\!C5.1.1.1.1 \quad O_{CARREL}(plan\_delivery(piece, origin\_org, dest\_org)$$
$$< do(deliver(piece, origin\_org, dest\_org)))$$
$$ONT\!:\!C5.1.1.1.1.1 \quad O_{CARREL}(plan\_delivery(organ, hospital_1, hospital_2)$$
$$< do(deliver(organ, hospital_1, hospital_2)))$$
$$ONT\!:\!C5.1.1.1.1.2 \quad O_{CARREL}(plan\_delivery(tissue, tissue\_bank, hospital)$$
$$< do(deliver(tissue, tissue\_bank, hospital)))$$
$$ONT\!:\!C5.1.1.2 \quad O_{board}(monitor\_logistics(piece, origin\_org, dest\_org))$$
$$ONT\!:\!C5.1.1.2.1 \quad O_{board}(monitor\_logistics(organ, hospital_1, hospital_2))$$
$$ONT\!:\!C5.1.1.2.2 \quad O_{board}(monitor\_logistics(tissue, tissue\_bank, hospital))$$

In this case the norms below are related to the obligation of organizing all the logistics involved in the delivery of a piece from the source organization to the one where the recipient is waiting for the piece. In this case such obligation has been divided in the one for the *CARREL* system to build the delivery plan and the one for the ONT's board, which should monitor any incidence that might happen during the delivery. In the case of *CARREL*'s obligations, a time constraint has also been introduced to express that the plan can be done at anytime but before the piece is sent. This allows several options for the planning such as having the plans pre-computed or building them during the assignation process.

Even though there is no written law, additional norms can be placed here in order to ensure that the *CARREL* system keeps the people of ONT's board updated of any (important) event that might happen, related either to the assignation or the delivery:

$$ONT:C8 \qquad O_{CARREL}(inform(board, events))$$
$$ONT:C8.1 \qquad O_{CARREL}(inform(board, distribution))$$
$$ONT:C8.2 \qquad O_{CARREL}(inform(board, logistics))$$

Influence of the Context of the E-Organization

At this level we have not found more norms that affect the organ and tissue allocation process, as most of the norms had a higher level of abstraction and were already introduced in the Abstract Level.

6.3.2 The Security Policy

The refinement process of the security policy is similar to the one presented for piece allocation. In this case, the initial set of norms contributing to the policy is the following:

$$ONT:C2 \qquad O_{ONT}(correct(distribution))$$
$$ONT:C4.1 \qquad O_{ONT}(anonymity(donor))$$
$$ONT:C4.2 \qquad O_{ONT}(anonymity(recipient))$$
$$ONT:C6 \qquad O_{ONT}(management(recipient\_waiting\_lists))$$
$$ONT:C7.1 \qquad O_{ONT}(record(donors))$$
$$ONT:C7.2 \qquad O_{ONT}(record(recipients))$$

Note that we have introduced again norm ONT:C2, which refers to the correctness of the distribution. While in the case of the allocation policies *correctness* is related to *efficiency*, in the case of security it is more related to concepts such as *security* and *accuracy* of data, as we will see in the following section.

Influence of the Context of the E-Organization

In the case of the security policy, the context of ONT (which is shared by *CARREL*) has a big influence. Because of norms ONT:C6, ONT:C7.1 and ONT:C7.2, the ONT is responsible of recording information about donors and recipients, information that is considered as *personal data* of the highest level by Spanish legislation (RD.994.99). Article 23 in RD.2070.99 explicitly says that the ONT should also follow the Spanish regulations on security of personal records (see Section 5.9 and Section A.3.2). Those regulations, interpreted for the case of ONT are the following:

$$RD.2070.99:23.1 \quad O_{ONT}(ensure\_accuracy(data)) \Rightarrow_{SpNHS} LOPD:4.3$$
$$RD.2070.99:23.2 \quad O_{ONT}(ensure\_security(data)) \Rightarrow_{SpNHS} LOPD:9$$
$$RD.2070.99:23.3 \quad O_{ONT}(notify(APD, new(datafile)) < do(use(datafile)))$$
$$\Rightarrow_{SpNHS} LOPD:44.3$$
$$RD.2070.99:23.4 \quad O_{ONT}(notify(APD, modification(datafile)) < do(use(datafile)))$$
$$\Rightarrow_{SpNHS} LOPD:20, LOPD:26.3$$
$$RD.2070.99:23.5 \quad O_{ONT}(notify(APD, first(transfer(data, org)) < do(send(data, org))))$$
$$\Rightarrow_{SpNHS} LOPD:27, LOPD:33$$

Note that these norms are interpretations made by SpNHS of the obligations imposed by Spanish Law regulations. We assume here that the norms we are linking (such as LOPD:4.3) are already expressed in Deontic Logic, so the $\Rightarrow_s$ operator links two expressions in Deontic Logic. The next step is to interpret those norms inside the ONT context:

$ONT:C2.2$ $O_{ONT}(ensure\_accuracy(patient, data)) \Rightarrow_{ONT} RD.2070.99:23.1$

$ONT:C2.3$ $O_{ONT}(ensure\_security(patient, data)) \Rightarrow_{ONT} RD.2070.99:23.2$

$ONT:C9$ $O_{ONT}(notify(APD, new(datafile) < do(use(datafile))))$
     $\Rightarrow_{ONT} RD.2070.99:23.3$

$ONT:C10$ $O_{ONT}(notify(APD, modification(datafile)) < do(use(datafile)))$
     $\Rightarrow_{ONT} RD.2070.99:23.4$

$ONT:C11$ $O_{ONT}(notify(APD, first(transfer(data, org)) < do(send(data, org))))$
     $\Rightarrow_{ONT} RD.2070.99:23.5$

In this case, the introduction of the norms has had two different effects:

- RD.2070.99:23.1 and RD.2070.99:23.2 are both related with the *correctness* (norm ONT:C2). In this case we refined (interpreted) a previous norm.

- RD.2070.99:23.3, RD.2070.99:23.4 and RD.2070.99:23.5 are related with no preexistent norm, so they are introduced as new norms.

Once we have introduced the effects of the context, we can refine the resulting norms in order to reduce their abstractness.

$ONT:C2.2.1$ $O_{hospital}(ensure\_accuracy(patient, data))$

$ONT:C2.2.2$ $O_{tissue\_bank}(ensure\_accuracy(donor, data))$

$ONT:C2.3.1$ $O_{CARREL}(ensure\_security(patient, data))$

$ONT:C2.3.2$ $O_{board}(ensure\_security(patient, data))$

$ONT:C2.3.3$ $O_{hospital}(ensure\_security(patient, data))$

$ONT:C2.3.4$ $O_{tissue\_bank}(ensure\_security(patient, data))$

This set of norms refines the norms related with accuracy and security of data. In this case such duties have been distributed among those entities that will manage such information: *CARREL*, the hospitals and the tissue banks.

$ONT:C4.1.1$ $O_{CARREL}(ensure\_anonymity(donor))$

$ONT:C4.1.2$ $O_{board}(ensure\_anonymity(donor))$

$ONT:C4.1.3$ $O_{hospital}(ensure\_anonymity(donor))$

$ONT:C4.1.4$ $O_{tissue\_bank}(ensure\_anonymity(donor))$

$ONT:C4.2.1$ $O_{CARREL}(ensure\_anonymity(recipient))$

$ONT:C4.2.2$ $O_{board}(ensure\_anonymity(recipient))$

$ONT:C4.2.3$ $O_{hospital}(ensure\_anonymity(recipient))$

In this case the norms related to anonymity have been refined. This case is similar to the one of the accuracy and security, but in this case we also enforce the board members of ONT to keep the anonymity.

$ONT:C6.1$ $\quad$ $O_{CARREL}(management(recipient\_waiting\_lists))$

$ONT:C6.1.1$ $\quad$ $O_{CARREL}(ensure\_security(recipient, recipient\_waiting\_lists))$
$\Rightarrow_{ONT} C2.2.1$

$ONT:C6.2$ $\quad$ $O_{hospital}(management(recipient\_waiting\_lists))$

$ONT:C6.2.2$ $\quad$ $O_{hospital}(ensure\_security(recipient, recipient\_waiting\_lists))$
$\Rightarrow_{ONT} C2.2.1$

$ONT:C7.1.1$ $\quad$ $O_{CARREL}(record(donors))$

$ONT:C7.1.1.1$ $\quad$ $O_{CARREL}(ensure\_security(donor, data)) \Rightarrow_{ONT} C2.2.1$

$ONT:C7.1.2$ $\quad$ $O_{hospital}(record(donors))$

$ONT:C7.1.2.1$ $\quad$ $O_{hospital}(ensure\_security(donor, data)) \Rightarrow_{ONT} C2.2.1$

$ONT:C7.1.3$ $\quad$ $O_{tissue\_bank}(record(donors))$

$ONT:C7.1.3.1$ $\quad$ $O_{tissue\_bank}(ensure\_security(donor, data)) \Rightarrow_{ONT} C2.2.1$

$ONT:C7.2.1$ $\quad$ $O_{CARREL}(record(recipients))$

$ONT:C7.2.1.1$ $\quad$ $O_{CARREL}(ensure\_security(recipient, data)) \Rightarrow_{ONT} C2.2.1$

$ONT:C7.2.1$ $\quad$ $O_{hospital}(record(recipients))$

$ONT:C7.2.1.1$ $\quad$ $O_{hospital}(ensure\_security(recipient, data)) \Rightarrow_{ONT} C2.2.1$

The norms above are related with the security issues of all the personal data the *CARREL* system manages (donors, recipients and recipient waiting lists). The first step in the refinement was to identify which are the entities that manage this kind of information (*CARREL*, hospitals and tissue banks) and then state that security should be ensured in the management of such information.

$ONT:C9.1$ $\quad$ $O_{board}(notify(APD, new(datafile) < do(use(datafile))))$

$ONT:C9.2$ $\quad$ $O_{hospital}(notify(APD, new(datafile) < do(use(datafile))))$

$ONT:C9.3$ $\quad$ $O_{tissue\_bank}(notify(APD, new(datafile) < do(use(datafile))))$

$ONT:C10.1$ $\quad$ $O_{board}(notify(APD, modification(datafile)) < do(use(datafile))))$

$ONT:C10.2$ $\quad$ $O_{hospital}(notify(APD, modification(datafile)) < do(use(datafile))))$

$ONT:C10.3$ $\quad$ $O_{tissue\_bank}(notify(APD, modification(datafile)) < do(use(datafile))))$

$ONT:C11.1$ $\quad$ $O_{board}(notify(APD, first(transfer(data, org)) < do(send(data, org))))$

$ONT:C11.2$ $\quad$ $O_{hospital}(notify(APD, first(transfer(data, org)) < do(send(data, org))))$

$ONT:C11.3$ $\quad$ $O_{tissue\_bank}(notify(APD, first(transfer(data, org)) < do(send(data, org))))$

Finally, this set of norms distributes the duties that any organization in Spain with personal data files has with the Spanish Data Protection Agency (APD). These duties include notifications of any new file with personal data, any modification of the logical structure of that file or any transfer that has not been previously approved by the APD. In this case such duties are not to be performed by a software application but by the organizations which are responsible of those files.

6.3.3 Summary: the Concrete Level in *CARREL*

Once all the norm refinement process is done, the next step is to identify which are the norms related to each of the entities identified so far (*CARREL*, ONT's board, hospitals and tissue banks). In the case of the hospitals and the tissue banks, the duties identified should be performed by members of those organizations or, in the case of having part of the process automated through a system, also by the system.[3] In order to be concise, we have not refined the norms involving those organizations as the purpose of this example is to find the norms to be followed by *CARREL*.

In the case of ONT's board and *CARREL*, the *e*-organization shares with the real organization the values, objectives and norms in the abstract level and most of the norms identified in the concrete level, as the separation of duties in those for *CARREL* and those for ONT's board is performed in the Concrete level.

From now on we will focus only in the refinement of *CARREL*. Following *definition 5.5*, the Concrete Level in *CARREL* can be described as follows:

$$conlevel = \langle c\_norms, ontology_{cl} \rangle$$

where, in this case,

- $c\_norms = \langle ONT{:}C1.1.1.2.1, ONT{:}C1.1.1.2.2, ONT{:}C1.2.1.1.1,$
  $ONT{:}C1.2.1.1.2, ONT{:}C2.1.1.1, ONT{:}C2.3.1, ONT{:}C3.2.1.1.1,$
  $ONT{:}C3.2.1.1.2, ONT{:}C3.2.1.2.1, ONT{:}C3.2.1.2.2, ONT{:}C4.1.1,$
  $ONT{:}C4.2.1, ONT{:}C5.1.1.1.1.1, ONT{:}C5.1.1.1.1.2, ONT{:}C6.1.1,$
  $ONT{:}C7.1.1.1, ONT{:}C7.2.1.1, ONT{:}C8.1, ONT{:}C8.2 \rangle$

- $ontology_{cl}$ is the ontology that defines all the terms and predicates that have appeared at this level in the refinement process.

With the definition of the abstract and concrete levels, the Normative System of *CARREL* is fully defined:

$$NS = \langle abslevel, conlevel \rangle$$

6.4   The Rule Level in *CARREL*

Once *CARREL*'s Normative System has been defined, the next phase is to define the Practical System. To do so a translation should be done from the normative dimension (which lacks of operational semantics) into the operational one.

In Section 5.4 we proposed to translate norms from Deontic Logic to Dynamic Logic. Summarizing, each norm should be translated to:

- a violation expression: following Meyer's reduction formulæ [4] from Deontic Logic to Dynamic Logic proposed in [138], the translation should apply the following rules:
  $$F(\alpha) \mapsto [\alpha] V$$
  $$P(\alpha) \equiv \neg F(\alpha) \mapsto <\alpha> \neg V$$
  $$O(\alpha) \equiv F(\neg \alpha) \mapsto [\neg \alpha] V$$

---

[3]In the case of hospitals we have designed the UCTx system (see **appendix B**). This system has to observe some of the norms we have identified for the hospitals, such as the ones about *data security* and *anonymity*.

[4]We already presented Meyer's reduction rules in Section 5.4.

- a precedence expression: in those cases where the norm expresses temporal relations among actions, such relation can be also expressed through the [ ] operator as follows:

$$O(\alpha < do(\beta)) \mapsto [\beta] \, done(\alpha)$$
$$O(\alpha < do(\beta)) \mapsto \neg done(\alpha) \rightarrow [\beta] \, V$$

The first reduction rule translates the temporal constraint of α being done before β with an expression in Dynamic Logic that states: *"once action β is performed, it should always be the case that action α has been done"*. The second reduction rule expresses the violation condition: *"if action α has not been done, once action β is performed it always is the case that violation V occurs"*.

We will translate each norm in the Concrete Level related to *CARREL* into one or more rules in Dynamic Logic. In order to make it easier to follow, we will split again the example in two: the organ and tissue allocation policies and the security policy.

6.4.1 The Organ and Tissue Allocation Policies

In the case of the organ and tissue allocation policies, the concrete norms related to the *CARREL* system are the following:

$ONT:C1.1.1.2.1$	$O_{CARREL}(get\_quality(organ, hospital)$ $< do(assign(organ, recipient)))$
$ONT:C1.1.1.2.2$	$O_{CARREL}(get\_quality(tissue, tissue\_bank)$ $< do(assign(tissue, recipient)))$
$ONT:C1.2.1.1.1$	$O_{CARREL}(ensure\_compatibility(organ, recipient)$ $< do(assign(organ, recipient)))$
$ONT:C1.2.1.1.2$	$O_{CARREL}(ensure\_compatibility(tissue, recipient)$ $< do(assign(tissue, recipient)))$
$ONT:C2.1.1.1$	$O_{CARREL}(seek\_time\_efficiency(distribution))$
$ONT:C3.2.1.1.1$	$F_{CARREL}(assign(organ, recipient) \wedge use(recipient, race))$
$ONT:C3.2.1.1.2$	$F_{CARREL}(assign(tissue, recipient) \wedge use(recipient, race))$
$ONT:C3.2.1.2.1$	$F_{CARREL}(assign(organ, recipient) \wedge use(recipient, sex))$
$ONT:C3.2.1.2.2$	$F_{CARREL}(assign(tissue, recipient) \wedge use(recipient, sex))$
$ONT:C5.1.1.1.1.1$	$O_{CARREL}(plan\_delivery(organ, hospital_1, hospital_2)$ $< do(deliver(organ, hospital_1, hospital_2)))$
$ONT:C5.1.1.1.1.2$	$O_{CARREL}(plan\_delivery(tissue, tissue\_bank, hospital)$ $< do(deliver(tissue, tissue\_bank, hospital)))$
$ONT:C8.1$	$O_{CARREL}(inform(board, distribution))$
$ONT:C8.2$	$O_{CARREL}(inform(board, logistics))$

Following these reduction rules, the translation of the norms above is as follows[5]:

$CRL:R1.1$ $[assign(organ, recipient)]done(get\_quality(organ, hospital))$
        $\mapsto_{CRL} ONT:C1.1.1.2.1$

$CRL:R1.2$ $\neg done(get\_quality(organ, hospital)$
        $\rightarrow [assign(organ, recipient)]CRL:V1 \mapsto_{CRL} ONT:C1.1.1.2.1$

$CRL:R2.1$ $[assign(tissue, recipient)]done(get\_quality(tissue, tissue\_bank))$
        $\mapsto_{CRL} ONT:C1.1.1.2.2$

$CRL:R2.2$ $\neg done(get\_quality(tissue, tissue\_bank))$
        $\rightarrow [assign(tissue, recipient)]CRL:V2 \mapsto_{CRL} ONT:C1.1.1.2.2$

$CRL:R3.1$ $[assign(organ, recipient)]done(ensure\_compatibility(organ, recipient))$
        $\mapsto_{CRL} ONT:C1.2.1.1.1$

$CRL:R3.2$ $\neg done(ensure\_compatibility(organ, recipient))$
        $\rightarrow [assign(organ, recipient)]CRL:V3 \mapsto_{CRL} ONT:C1.2.1.1.1$

$CRL:R4.1$ $[assign(tissue, recipient)]done(ensure\_compatibility(tissue, recipient))$
        $\mapsto_{CRL} ONT:C1.2.1.1.2$

$CRL:R4.2$ $\neg done(ensure\_compatibility(tissue, recipient))$
        $\rightarrow [assign(tissue, recipient)]CRL:V4 \mapsto_{CRL} ONT:C1.2.1.1.2$

$CRL:R5$ $[\neg seek\_time\_efficiency(distribution)]CRL:V5 \mapsto_{CRL} ONT:C2.1.1.1$

$CRL:R6$ $[assign(organ, recipient) \wedge use(recipient, race)]CRL:V6$
        $\mapsto_{CRL} ONT:C3.2.1.1.1$

$CRL:R7$ $[assign(tissue, recipient) \wedge use(recipient, race)]CRL:V7$
        $\mapsto_{CRL} ONT:C3.2.1.1.2$

$CRL:R8$ $[assign(organ, recipient) \wedge use(recipient, sex)]CRL:V8$
        $\mapsto_{CRL} ONT:C3.2.1.2.1$

$CRL:R9$ $[assign(tissue, recipient) \wedge use(recipient, sex)]CRL:V9$
        $\mapsto_{CRL} ONT:C3.2.1.2.2$

$CRL:R10.1$ $[deliver(organ, hospital_1, hospital_2)]done(plan\_delivery(organ, hospital_1,$
        $hospital_2)) \mapsto_{CRL} ONT:C5.1.1.1.1.1$

$CRL:R10.2$ $\neg done(plan\_delivery(organ, hospital_1, hospital_2))$
        $\rightarrow [deliver(organ, hospital_1, hospital_2)]CRL:V10$
        $\mapsto_{CRL} ONT:C5.1.1.1.1.1$

$CRL:R11.1$ $[deliver(tissue, tissue\_bank, hospital)]done(plan\_delivery(tissue, tissue\_bank,$
        $hospital)) \mapsto_{CRL} ONT:C5.1.1.1.1.2$

$CRL:R11.2$ $\neg done(plan\_delivery(tissue, tissue\_bank, hospital))$
        $\rightarrow [deliver(tissue, tissue\_bank, hospital)]CRL:V11$
        $\mapsto_{CRL} ONT:C5.1.1.1.1.2$

$CRL:R12$ $[\neg inform(board, distribution)]CRL:V12 \mapsto_{CRL} ONT:C8.1$

$CRL:R13$ $[\neg inform(board, logistics)]CRL:V13 \mapsto_{CRL} ONT:C8.2$

We will talk more about violations and their related sanctions in Section 6.4.3.

Apart from the translation of the obligations that were assigned to *CARREL*, at this level the designer should also check if the control of some of the obligations of the other

---

[5]Notation: All violations are identified by a code following the format $CRL:Vx$.

actors (board, hospitals, tissue_banks) should be carried out by the *CARREL* system. We should only check the obligations of those roles which *CARREL* has a power relation to (i.e., it is *empowered* to, among other things, check those roles' behaviour). As seen in Section 6.2, those roles are *hospital* and *tissue_bank*). In this case there are two obligations of these two roles to be checked:

$$ONT:C1.1.1.1.1 \quad O_{hospital}(ensure\_quality(organ))$$
$$ONT:C1.1.1.1.2 \quad O_{tissue\_bank}(ensure\_quality(tissue))$$

Although the *CARREL* system cannot really ensure that the quality control has been properly performed, it can check that hospitals and tissue banks send, along with the data about the organ and tissue, some fields ensuring that the quality tests were done.

$$CRL:R14.1 \quad [\neg hospital.ensure\_quality(organ)]CRL:V14 \mapsto_{CRL} ONT:C1.1.1.1.1$$
$$CRL:R15.1 \quad [\neg tissue\_bank.ensure\_quality(tissue)]CRL:V15 \mapsto_{CRL} ONT:C1.1.1.1.2$$

Note that, in order to specify that the referred procedures are outside the *CARREL* system, we have added to those actions the entity which should perform them. The way those external procedures are checked by *CARREL* is expressed as follows:

$$CRL:R14.2 \quad get\_quality(organ, hospital) \rightarrow_{CRL} hospital.ensure\_quality(organ)$$
$$CRL:R15.2 \quad get\_quality(tissue, tissue\_bank) \rightarrow_{CRL} tissue\_bank.ensure\_quality(tissue)$$

These rules, in fact, create also links with other rules previously defined. For instance, rule CRL:R14.2 links rules CRL:R14.1 with CRL:R1.1 and CRL:R1.2, as it refers to the *get_quality* procedure. Therefore, violations CRL:V1 and CRL:V14 are more or less related. However, those are two different violations that should be treated in a different way:

- CRL:V14 is a violation of a hospital that has not certified that the proper quality tests were performed. As we will see in Section 6.4.3, the sanction imposed in this case is that the hospital should send again the data with the proper information about the quality tests.

- CRL:V1 is a violation of the *CARREL* system, if ever assigns an organ without checking first that the hospital sent the proper information about the quality tests.

Influence of the Context of the E-Organization

At this point we should also consider rules that context may impose on the allocation process. In this case the context of ONT imposes no rules, as it gives ONT the full power to rule in the allocation process.

There are some rules ONT has defined and are important to be introduced here. One of them is about the precedence of urgency_0 patients over any other patient (see Section 2.3.2):

$$(compatible(patient_1, piece) \wedge compatible(patient_2, piece)$$
$$\wedge urgency\_0(patient_1)) \Rightarrow assign(patient_1, piece)$$

With such a rule we express that, given two patients that are compatible with a given piece, if one of them has the urgency_0 level then the piece should be assigned to the urgency_0 patient. We can express this rule as follows:

$CRL:R16.1$ $[assign(organ, patient)]done(check\_if\_urgency\_0(organ)) \Rightarrow_{CRL} ONT:R1$

$CRL:R16.2$ $\neg done(check\_if\_urgency\_0(organ)) \rightarrow [assign(organ, patient)]CRL:V16$
            $\Rightarrow_{CRL} ONT:R1$

$CRL:R17.1$ $[assign(tissue, patient)]done(check\_if\_urgency\_0(tissue)) \Rightarrow_{CRL} ONT:R1$

$CRL:R17.2$ $\neg done(check\_if\_urgency\_0(tissue)) \rightarrow [assign(tissue, patient)]CRL:V17$
            $\Rightarrow_{CRL} ONT:R1$

Note here the use of the *counts-as* operator to express that these rule are interpretations of the above-mentioned ONT's rule (which we have tagged as ONT:R1).[6]

A unwritten rule that ONT is trying to implement is that hospitals should return information about the result of the transplantation from each piece, in order to record the results of the implant and also to allow ONT to detect any weak point in the allocation process to be amended.[7] We can express such a rule as follows:

$CRL:R18.1$ $[deliver(organ, hospital_1, hospital_2)](is\_time(t) \rightarrow$
        $[wait\_for(is\_time(t + 1\ day))]done(hospital_2.send\_result\_transplant(recipient)))$
        $\Rightarrow_{CRL} ONT:R2$

$CRL:R18.2$ $(done(deliver(organ, hospital_1, hospital_2)) \wedge is\_time(t)$
        $\wedge \neg done(hospital_2.send\_result\_transplant(recipient)))$
        $\rightarrow [wait\_for(is\_time(t + 1\ day))]CRL:V18.1 \Rightarrow_{CRL} ONT:R2$

The first rule expresses the following: *"once the organ is sent, if* t *is the current time then when time* t + 1 *day has passed it always should be that the hospital has sent the data about the transplantation".* The second is the violation rule, which expresses the following: *"if the organ has been sent at time* t *and the hospital has not sent the data about the transplantation, once time* (t + 1 day) *passes the violation CRL:V18.1 is triggered".* The same structure can be used to express the evolution reports three months later and a year later, and to express the same rules in the case of tissues:

$CRL:R18.3$ $[deliver(organ, hospital_1, hospital_2)](is\_time(t) \rightarrow [wait\_for($
            $is\_time(t + 3\ months))]done(hospital_2.send\_evolution\_3months(recipient)))$
            $\Rightarrow_{CRL} ONT:R2$

$CRL:R18.4$ $(done(deliver(organ, hospital_1, hospital_2)) \wedge is\_time(t)$
            $\wedge \neg done(hospital_2.send\_evolution\_3months(recipient)))$
            $\rightarrow [wait\_for(is\_time(t + 3\ months))]CRL:V18.2 \Rightarrow_{CRL} ONT:R2$

$CRL:R18.5$ $[deliver(organ, hospital_1, hospital_2)](is\_time(t) \rightarrow [wait\_for($
            $is\_time(t + 1\ year))]done(hospital_2.send\_evolution\_1year(recipient)))$
            $\Rightarrow_{CRL} ONT:R2$

---

[6] In the examples of this section we will see again the use of the $\Rightarrow_s$ operator referring to external rules. In all those cases we assume that those external rules are also formalized in terms of Dynamic Logic.

[7] Now a days, ONT only keeps a nation-wide registry about liver transplantation. On the other hand, OCATT already keeps registries about heart, liver, kidney and stem cells.

$CRL:R18.6$ $(done(deliver(organ, hospital_1, hospital_2)) \land is\_time(t)$
        $\land \neg done(hospital_2.send\_evolution\_1year(recipient)))$
        $\rightarrow [wait\_for(is\_time(t + 1\ year))]CRL:V18.3 \Rightarrow_{CRL} ONT:R2$

$CRL:R18.7$ $[deliver(tissue, tissue\_bank, hospital)](is\_time(t) \rightarrow [wait\_for($
        $is\_time(t + 1\ day))]done(hospital.send\_result\_transplant(recipient)))$
        $\Rightarrow_{CRL} ONT:R2$

$CRL:R18.8$ $(done(deliver(tissue, tissue\_bank, hospital)) \land is\_time(t)$
        $\land \neg done(hospital.send\_result\_transplant(recipient)))$
        $\rightarrow [wait\_for(is\_time(t + 1\ day))]CRL:V18.4 \Rightarrow_{CRL} ONT:R2$

$CRL:R18.9$ $[deliver(tissue, tissue\_bank, hospital)](is\_time(t) \rightarrow [wait\_for($
        $is\_time(t + 3\ months))]done(hospital.send\_evolution\_3months(recipient)))$
        $\Rightarrow_{CRL} ONT:R2$

$CRL:R18.10$ $(done(deliver(tissue, tissue\_bank, hospital)) \land is\_time(t)$
        $\land \neg done(hospital.send\_evolution\_3months(recipient)))$
        $\rightarrow [wait\_for(is\_time(t + 3\ months))]CRL:V18.5 \Rightarrow_{CRL} ONT:R2$

$CRL:R18.11$ $[deliver(tissue, tissue\_bank, hospital)](is\_time(t) \rightarrow [wait\_for($
        $is\_time(t + 1\ year))]done(hospital.send\_evolution\_1year(recipient)))$
        $\Rightarrow_{CRL} ONT:R2$

$CRL:R18.12$ $(done(deliver(tissue, tissue\_bank, hospital)) \land is\_time(t)$
        $\land \neg done(hospital.send\_evolution\_1year(recipient)))$
        $\rightarrow [wait\_for(is\_time(t + 1\ year))]CRL:V18.6 \Rightarrow_{CRL} ONT:R2$

Checking for Completeness

Once we get the set of rules coming from the Normative System and the ones coming from the surrounding context, we should add some more rules in order to make the model complete.

The addition of rules should address two aspects:

- the completeness of the precedence relations between actions: usually the norms and the context impose some precedence restrictions but not all of them (e.g., the *common sense* precedence relations are skipped). Figure 6.6 sketches the precedence relations among actions (defined in rules CRL:R1.1, CRL:R2.1, CRL:R3.1, CRL:R4.1, CRL:R10.1, and CRL:R11.1). It also identifies a missing relation to state that assignation of pieces should be done before sending them to the destination. Therefore, we should add the following rules:

     $CRL:R19.1$ $[deliver(organ, hospital_1, hospital_2)]done(assign(organ, recipient))$
     $CRL:R19.2$ $[deliver(tissue, tissue\_bank, hospital)]done(assign(tissue, recipient))$

- the refinement of actions and predicates: Domain Knowledge should also be introduced at this phase in the form of rules that fully specify some predicates and actions. We will not present here all the rules that emerge from this step, but some examples are the rules refining the $ensure\_quality$ and $ensure\_compatibility$ for kidneys (see Section 5.4).

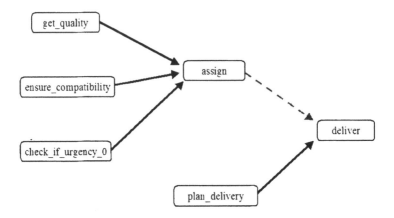

Figure 6.6: Sketch of the precedence relations among actions in the (organ and tissue) allocation policies.

6.4.2 The Security Policy

In the case of the security policy, the concrete norms related to the *CARREL* system are the following:

$$ONT:C4.1.1 \quad O_{CARREL}(ensure\_anonymity(donor))$$
$$ONT:C4.2.1 \quad O_{CARREL}(ensure\_anonymity(recipient))$$
$$ONT:C6.1.1 \quad O_{CARREL}(ensure\_security(recipient,\ recipient\_waiting\_lists))$$
$$ONT:C7.1.1 \quad O_{CARREL}(ensure\_security(donor,\ data))$$
$$ONT:C7.2.1 \quad O_{CARREL}(ensure\_security(recipient,\ data))$$

Once the concrete norms are identified, they should be translated following the reduction rules described in Section 6.4.1. The resulting set of rules is the following:

$$CRL:R20 \quad [\neg ensure\_anonymity(donor)]CRL:V19 \mapsto_{CRL} ONT:C4.1.1$$
$$CRL:R21 \quad [\neg ensure\_anonymity(recipient)]CRL:V20 \mapsto_{CRL} ONT:C4.2.1$$
$$CRL:R22 \quad [\neg ensure\_security(recipient,\ recipient\_waiting\_lists)]CRL:V21$$
$$\mapsto_{CRL} ONT:C6.1.1$$
$$CRL:R23 \quad [\neg ensure\_security(donor,donor\_data)]CRL:V22 \mapsto_{CRL} ONT:C7.1.1$$
$$CRL:R24 \quad [\neg ensure\_security(recipient,\ recipient\_data)]CRL:V23$$
$$\mapsto_{CRL} ONT:C7.2.1$$

As we did in the case of the organ and tissue allocation policies, once the translation is done for the obligations *CARREL* was assigned, we should then check if some of the obligations assigned to other roles may be controlled by *CARREL*. In this case there are several of such obligations:

$$ONT:C2.2.1 \quad O_{hospital}(ensure\_accuracy(patient, data))$$
$$ONT:C2.2.2 \quad O_{tissue\_bank}(ensure\_accuracy(donor, data))$$
$$ONT:C7.1.2.1 \quad O_{hospital}(ensure\_security(donor, data))$$
$$ONT:C7.1.3.1 \quad O_{tissue\_bank}(ensure\_security(donor, data))$$
$$ONT:C7.2.1.1 \quad O_{hospital}(ensure\_security(recipient, data))$$

All these obligations are related to accuracy and security of the data about patients managed by hospitals and tissue banks. Although *CARREL* cannot directly ensure that all these obligations are fulfilled when information is managed and exchanged inside those organizations, *CARREL* may check if the obligation is met when they send data to the *CARREL* system. Therefore, we will translate those obligations in rules as follows:

$$CRL:R25.1 \quad [\neg hospital.ensure\_accuracy(donor, data)]CRL:V24 \mapsto ONT:C2.2.1$$
$$CRL:R25.2 \quad [\neg hospital.ensure\_accuracy(recipient, data)]CRL:V25 \mapsto ONT:C2.2.1$$
$$CRL:R26 \quad [\neg tissue\_bank.ensure\_accuracy(donor, data)]CRL:V26 \mapsto ONT:C2.2.2$$
$$CRL:R27.1 \quad [\neg hospital.ensure\_security(donor, data)]CRL:V27 \mapsto ONT:C7.1.2.1$$
$$CRL:R27.2 \quad [\neg hospital.ensure\_security(recipient, data)]CRL:V28 \mapsto ONT:C7.2.1.1$$
$$CRL:R28 \quad [\neg tissue\_bank.ensure\_security(donor, data)]CRL:V29 \mapsto ONT:C7.1.3.1$$

Next step is to define the link of these predicates (which refer to external procedures) with the ones inside *CARREL*:

$$CRL:R29.1 \quad input(data, hospital) \rightarrow_{CRL} hospital.ensure\_accuracy(donor, data)$$
$$CRL:R29.2 \quad input(data, hospital) \rightarrow_{CRL} hospital.ensure\_accuracy(recipient, data)$$
$$CRL:R29.3 \quad input(data, tissue\_bank) \rightarrow_{CRL} tissue\_bank.ensure\_accuracy(donor, data)$$
$$CRL:R30.1 \quad input(data, hospital) \rightarrow_{CRL} hospital.ensure\_security(donor, data)$$
$$CRL:R30.2 \quad input(data, hospital) \rightarrow_{CRL} hospital.ensure\_security(recipient, data)$$
$$CRL:R30.3 \quad input(data, tissue\_bank) \rightarrow_{CRL} tissue\_bank.ensure\_security(donor, data)$$

The rules above state that the reception by the *CARREL* system of any data from a hospital or a tissue bank is related with ensuring accuracy, anonymity and security of the received data.

Influence of the Context of the E-Organization

At this level there is a big influence of the context in the security policy. Security measures have been defined by the European Commission (Article 9.2 in R.97.5) to be applied by organizations managing personal information. Some of these measures have been extended later by Spanish Law in RD.994.99 and Article 9.2 in R.97.5. The measures [8] are the following:

1. *User identification*: R.97.5:9.2d and RD.994.99:18 state that identification mechanisms should be introduced in order to ensure that only authorised persons and

---

[8]These security measures are described in Section A.5.

institutions can access to system. This rule clearly should be part of the rules in *CARREL*. Therefore, we can express the rule as follows:[9]

$RD.994.99 : 18.1 \quad [access(x, system)]done(check\_identity(x)) \Rightarrow_{SpLaw} R.97.5 : 9.2d$

$RD.994.99 : 18.2 \quad \neg done(check\_identity(x)) \rightarrow [access(x, system)]V$
$\qquad\qquad\qquad \Rightarrow_{SpLaw} R.97.5 : 9.2d$

2. *Facility access control*: R.97.5:9.2a and RD.994.99:19 state that no unauthorized person must be able to access to the facilities where personal data are stored or processed. This rule is clearly outside the scope of *CARREL*, as it is a duty for the staff at ONT.

3. *Data media control*: R.97.5:9.2b states that unauthorized people cannot copy, alter or take away the data. This can be expressed as follows:

$R.97.5 : 9.2b.1 \quad [store(data, disc)]done(avoid\_possible\_attack(data))$

$R.97.5 : 9.2b.2 \quad \neg done(avoid\_possible\_attack(data)) \rightarrow [store(data, disc)]V$

4. *Memory and telematic transmissions*: R.97.5:9.2c, R.97.5:9.2h and RD.994.99:26 state that security of data is to be ensured also when the information is at the computer's memory or when is sent through a network. In this case we can express the rules as follows:

$R.97.5 : 9.2c.1 \quad [store(data, memory)]done(avoid\_possible\_attack(data))$

$R.97.5 : 9.2c.2 \quad \neg done(avoid\_possible\_attack(data)) \rightarrow [store(data, memory)]V$

$RD.994.99 : 26.1 \quad [send(data, network)]done(avoid\_possible\_attack(data))$
$\qquad\qquad\qquad \Rightarrow_{SpLaw} R.97.5 : 9.2h$

$RD.994.99 : 26.2 \quad \neg done(avoid\_possible\_attack(data)) \rightarrow [send(data, network)]V$
$\qquad\qquad\qquad \Rightarrow_{SpLaw} R.97.5 : 9.2h$

5. *Usage control*: RD.994.99:12.2 states that data must be protected against unauthorized processing. Therefore, the rights of the user should be checked before the user accesses some data. We can express this as follows:

$RD.994.99 : 12.2.1 \quad [input(x, data)]done(check\_access\_rights(x, data))$

$RD.994.99 : 12.2.2 \quad \neg done(check\_access\_rights(x, data)) \rightarrow [input(x, data)]V$

$RD.994.99 : 12.2.3 \quad [query(x, data)]done(check\_access\_rights(x, data))$

$RD.994.99 : 12.2.4 \quad \neg done(check\_access\_rights(x, data)) \rightarrow [query(x, data)]V$

6. *System design*: R.97.5:9.2e states that the design of data structures and procedures should separate identifiers and data related to person identity from the rest of data.

---

[9]Note that in the case of the violation rule, we only introduce a undefined violation V, which will be then defined later when we introduce all the rules in *CARREL*'s specification.

We can express this as follows:

$R.97.5 : 9.2e.1$ $[store(data, memory)]done(dissociate(patient\_id, data)$

$R.97.5 : 9.2e.2$ $\neg done(dissociate(patient\_id, data)) \rightarrow [store(data, memory)]V$

$R.97.5 : 9.2e.3$ $[store(data, disc)]done(dissociate(patient\_id, data)$

$R.97.5 : 9.2e.4$ $\neg done(dissociate(patient\_id, data)) \rightarrow [store(data, disc)]V$

7. *Data loss protection*: RD.994.99:14 states that organizations should protect their data from accidental or illegal destruction or loss. In this case this is a duty for system administrators and not for a system such as *CARREL*.

8. *Data recovery*: R.97.5:9.2i and RD.994.99:14 state that organizations should also do backups and define procedures for recovery in case of partial or full data destruction or loss. Again this is something for system administrators to be dealt with. [10]

9. *Access and data input logging*: R.97.5:9.2g and RD.994.99:24 state that the system must record (into log files) when and who accessed the system and which information has been entered:[11]

$RD.994.99 : 24.1$ $[access(x, system)](is\_time(t) \rightarrow [wait\_for($
$is\_time(t + n\ min))]done(record(info\_access(x, related\_info), log))))$
$\Rightarrow_{SpLaw} R.97.5 : 9.2g$

$RD.994.99 : 24.2$ $(done(access(x, system)) \wedge is\_time(t)$
$\wedge done(record(info\_access(x, related\_info), log))))$
$\rightarrow [wait\_for(is\_time(t + n\ min))]V \Rightarrow_{SpLaw} R.97.5 : 9.2g$

$RD.994.99 : 24.3$ $[input(x, data)](is\_time(t) \rightarrow [wait\_for($
$is\_time(t + n\ min))]done(record(info\_input(x, related\_info), log))))$
$\Rightarrow_{SpLaw} R.97.5 : 9.2g$

$RD.994.99 : 24.4$ $(done(input(x, data)) \wedge is\_time(t)$
$\wedge done(record(info\_input(x, related\_info), log))))$
$\rightarrow [wait\_for(is\_time(t + n\ min))]V \Rightarrow_{SpLaw} R.97.5 : 9.2g$

$RD.994.99 : 24.5$ $[query(x, data)](is\_time(t) \rightarrow [wait\_for($
$is\_time(t + n\ min))]done(record(info\_query(x, related\_info), log))))$
$\Rightarrow_{SpLaw} R.97.5 : 9.2g$

$RD.994.99 : 24.6$ $(done(query(x, data)) \wedge is\_time(t)$
$\wedge done(record(info\_query(x, related\_info), log))))$
$\rightarrow [wait\_for(is\_time(t + n\ min))]V \Rightarrow_{SpLaw} R.97.5 : 9.2g$

10. *Incidence log*: RD.994.99:21 states that an incident log should be maintained for any major incident (e.g. unauthorized access attempts, system failures leading to a

---

[10] At this point the Background Knowledge can be used to check if some of this duties can be taken by, e.g., database backup and recovery policies.

[11] Note that we use again the Dynamic Logic formulæ structure we used back in rules CRL:R18.1 to CRL:R18.12, in order to express here that a given action (*record*) has to be performed shortly after another action (*access, input, query*). As Spanish Law does not impose a concrete time limit for record to occur after the event, we have placed here a undefined time n, which we will later refine).

restore process from backups). This can be expressed as follows:

$RD.994.99 : 21.1$ $[incident(system)] (is\_time(t) \rightarrow [wait\_for($
$is\_time(t + n\ min))] done(record(info\_incident(related\_info), log))))$

$RD.994.99 : 21.2$ $(done(incident(system)) \wedge is\_time(t)$
$\wedge done(record(info\_incident(related\_info), log))))$
$\rightarrow [wait\_for(is\_time(t + n\ min))] V$

Once we have identified the rules imposed by the context of ONT, the next step is to introduce them in the rule set. Some of these rules from the context make reference to predicates in *CARREL*'s existing rules. In this case the rule of the context will refine existing rules. Some of the rules coming from the context are not related to existing predicates. In this case, as they introduce new restrictions, the rules are introduced directly into the rule set. The result of this process is described below.

$CRL : R20.1$ $[store(data, memory)] done(dissociate(donor\_id, data))$
$\Rightarrow_{CRL} R.97.5 : 9.2e.1$

$CRL : R20.2$ $\neg done(dissociate(donor\_id, data)) \rightarrow [store(data, memory)] CRL : V19.1$
$\Rightarrow_{CRL} R.97.5 : 9.2e.2$

$CRL : R20.3$ $[store(data, disc)] done(dissociate(donor\_id, data)) \Rightarrow_{CRL} R.97.5 : 9.2e.3$

$CRL : R20.4$ $\neg done(dissociate(donor\_id, data)) \rightarrow [store(data, disc)] CRL : V19.2$
$\Rightarrow_{CRL} R.97.5 : 9.2e.4$

$CRL : R21.1$ $[store(data, memory)] done(dissociate(recipient\_id, data))$
$\Rightarrow_{CRL} R.97.5 : 9.2e.1$

$CRL : R21.2$ $\neg done(dissociate(recipient\_id, data)) \rightarrow [store(data, memory)] CRL : V20.1$
$\Rightarrow_{CRL} R.97.5 : 9.2e.2$

$CRL : R21.3$ $[store(data, disc)] done(dissociate(recipient\_id, data))$
$\Rightarrow_{CRL} R.97.5 : 9.2e.3$

$CRL : R21.4$ $\neg done(dissociate(recipient\_id, data)) \rightarrow [store(data, disc)] CRL : V20.2$
$\Rightarrow_{CRL} R.97.5 : 9.2e.4$

$CRL : R22.1$ $[store(recipient\_waiting\_lists, disc)] done($
$avoid\_possible\_attack(recipient\_waiting\_lists)) \Rightarrow_{CRL} R.97.5 : 9.2b.1$

$CRL : R22.2$ $\neg done(avoid\_possible\_attack(recipient\_waiting\_lists))$
$\rightarrow [store(recipient\_waiting\_lists, disc)] CRL : V21.1$
$\Rightarrow_{CRL} R.97.5 : 9.2b.2$

$CRL : R22.3$ $[store(recipient\_waiting\_lists, memory)] done($
$avoid\_possible\_attack(recipient\_waiting\_lists)) \Rightarrow_{CRL} R.97.5 : 9.2c.1$

$CRL : R22.4$ $\neg done(avoid\_possible\_attack(recipient\_waiting\_lists))$
$\rightarrow [store(recipient\_waiting\_lists, memory)] CRL : V21.2$
$\Rightarrow_{CRL} R.97.5 : 9.2c.2$

$CRL : R22.5$ $[send(recipient\_waiting\_lists, network)] done($
$avoid\_possible\_attack(recipient\_waiting\_lists)) \Rightarrow_{CRL} RD.994.99 : 26.1$

$CRL : R22.6$ $\neg done(avoid\_possible\_attack(recipient\_waiting\_lists))$
$\rightarrow [send(recipient\_waiting\_lists, network)] CRL : V21.3$
$\Rightarrow_{CRL} RD.994.99 : 26.2$

$CRL : R23.1$ $[store(donor\_data, disc)]done(avoid\_possible\_attack(donor\_data))$
$\Rightarrow_{CRL} R.97.5 : 9.2b.1$

$CRL : R23.2$ $\neg done(avoid\_possible\_attack(donor\_data))$
$\rightarrow [store(donor\_data, disc)]CRL : V22.1 \Rightarrow_{CRL} R.97.5 : 9.2b.2$

$CRL : R23.3$ $[store(donor\_data, memory)]done(avoid\_possible\_attack(donor\_data))$
$\Rightarrow_{CRL} R.97.5 : 9.2c.1$

$CRL : R23.4$ $\neg done(avoid\_possible\_attack(donor\_data))$
$\rightarrow [store(donor\_data, memory)]CRL : V22.2 \Rightarrow_{CRL} R.97.5 : 9.2c.2$

$CRL : R23.5$ $[send(donor\_data, network)]done(avoid\_possible\_attack(donor\_data))$
$\Rightarrow_{CRL} RD.994.99 : 26.1$

$CRL : R23.6$ $\neg done(avoid\_possible\_attack(donor\_data))$
$\rightarrow [send(donor\_data, network)]CRL : V22.3 \Rightarrow_{CRL} RD.994.99 : 26.2$

$CRL : R24.1$ $[store(recipient\_data, disc)]done(avoid\_possible\_attack(recipient\_data))$
$\Rightarrow_{CRL} R.97.5 : 9.2b.1$

$CRL : R24.2$ $\neg done(avoid\_possible\_attack(recipient\_data))$
$\rightarrow [store(recipient\_data, disc)]CRL : V23.1$
$\Rightarrow_{CRL} R.97.5 : 9.2b.2$

$CRL : R24.3$ $[store(recipient\_data, memory)]done(avoid\_possible\_attack(recipient\_data))$
$\Rightarrow_{CRL} R.97.5 : 9.2c.1$

$CRL : R24.4$ $\neg done(avoid\_possible\_attack(recipient\_data))$
$\rightarrow [store(recipient\_data, memory)]CRL : V23.2$
$\Rightarrow_{CRL} R.97.5 : 9.2c.2$

$CRL : R24.5$ $[send(recipient\_data, network)]done(avoid\_possible\_attack(recipient\_data))$
$\Rightarrow_{CRL} RD.994.99 : 26.1$

$CRL : R24.6$ $\neg done(avoid\_possible\_attack(recipient\_data))$
$\rightarrow [send(recipient\_data, network)]CRL : V23.3$
$\Rightarrow_{CRL} RD.994.99 : 26.2$

In the rule set listed above we can see how some of the rules imposed by the surrounding context (about dissociation and attack avoidance) have refined existing rules about anonymity and security (rules CRL:R20 to CRL:R24) which apply to *CARREL*. This refinement imposes some steps to be followed (e.g., ensuring security is avoiding possible attacks before the information is sent), and introduces in the rule set 1) rules in Dynamic Logic describing the precedence among steps, and 2) refinement of existing violation rules.

$CRL : R27.1.1$ $[hospital.send(donor\_data, CARREL)]done($
$hospital.avoid\_possible\_attack(donor\_data)) \Rightarrow_{CRL} RD.994.99 : 26.1$

$CRL : R27.1.2$ $\neg done(hospital.avoid\_possible\_attack(donor\_data))$
$\rightarrow [hospital.send(donor\_data, CARREL)]CRL : V27$
$\Rightarrow_{CRL} RD.994.99 : 26.2$

$CRL : R27.2.1$ $[hospital.send(recipient\_data, CARREL)]done($
$hospital.avoid\_possible\_attack(recipient\_data)) \Rightarrow_{CRL} RD.994.99 : 26.1$

$CRL:R27.2.2$ $\neg done(hospital.avoid\_possible\_attack(recipient\_data))$
$\rightarrow [hospital.send(recipient\_data, CARREL)]CRL:V28$
$\Rightarrow_{CRL} RD.994.99 : 26.2$

$CRL:R28.1$ $[tissue\_bank.send(donor\_data, CARREL)]done($
$tissue\_bank.avoid\_possible\_attack(donor\_data)) \Rightarrow_{CRL} RD.994.99 : 26.1$

$CRL:R28.2$ $\neg done(tissue\_bank.avoid\_possible\_attack(donor\_data))$
$\rightarrow [tissue\_bank.send(donor\_data, CARREL)]CRL:V29$
$\Rightarrow_{CRL} RD.994.99 : 26.2$

In this second rule set we can see how the rules imposed by the context of *CAR-REL* about attack avoidance have refined the existing rules about security (CRL:R27.1 to CRL:R28) applying to hospitals and tissue banks. The process is similar as for rules CRL:R20 to CRL:R24, as it introduces rules describing the precedence among steps, and refinement of the existing violation rules.

$CRL:R31.1$ $[access(agent, CARREL)]done(check\_identity(agent))$
$\Rightarrow_{CRL} RD.994.99 : 18.1$

$CRL:R31.2$ $\neg done(check\_identity(agent)) \rightarrow [access(agent, CARREL)]CRL:V30$
$\Rightarrow_{CRL} RD.994.99 : 18.2$

$CRL:R32.1$ $[input(agent, data)]done(check\_access\_rights(agent, data))$
$\Rightarrow_{CRL} RD.994.99 : 12.2.1$

$CRL:R32.2$ $\neg done(check\_access\_rights(agent, data)) \rightarrow [input(agent, data)]CRL:V31$
$\Rightarrow_{CRL} RD.994.99 : 12.2.2$

$CRL:R33.1$ $[query(agent, data)]done(check\_access\_rights(agent, data))$
$\Rightarrow_{CRL} RD.994.99 : 12.2.3$

$CRL:R33.2$ $\neg done(check\_access\_rights(agent, data)) \rightarrow [query(agent, data)]CRL:V32$
$\Rightarrow_{CRL} RD.994.99 : 12.2.4$

$CRL:R34.1$ $[access(agent, CARREL)](is\_time(t) \rightarrow [wait\_for($
$is\_time(t + 1\ min))]done(record(info\_access(agent, related\_info), log))))$
$\Rightarrow_{SpLaw} RD.994.99 : 24.1$

$CRL:R34.2$ $(done(access(agent, CARREL)) \wedge is\_time(t)$
$\wedge \neg done(record(info\_access(agent, related\_info), log))))$
$\rightarrow [wait\_for(is\_time(t + 1\ min))]CRL:V33 \Rightarrow_{SpLaw} RD.994.99 : 24.2$

$CRL:R35.1$ $[input(agent, data)](is\_time(t) \rightarrow [wait\_for($
$is\_time(t + 1\ min))]done(record(info\_input(x, related\_info), log))))$
$\Rightarrow_{SpLaw} RD.994.99 : 24.3$

$CRL:R35.2$ $(done(input(agent, data)) \wedge is\_time(t)$
$\wedge \neg done(record(info\_input(agent, related\_info), log))))$
$\rightarrow [wait\_for(is\_time(t + 1\ min))]CRL:V34 \Rightarrow_{SpLaw} RD.994.99 : 24.4$

$CRL:R36.1$ $[query(agent, data)](is\_time(t) \rightarrow [wait\_for($
$is\_time(t + 1\ min))]done(record(info\_query(x, related\_info), log))))$
$\Rightarrow_{SpLaw} RD.994.99 : 24.5$

$CRL:R36.2$ $(done(query(agent, data)) \wedge is\_time(t)$
$\wedge \neg done(record(info\_query(agent, related\_info), log))))$
$\rightarrow [wait\_for(is\_time(t + 1\ min))]CRL:V35 \Rightarrow_{SpLaw} RD.994.99 : 24.6$

$CRL: R37.1$ $[incident(CARREL)](is\_time(t) \rightarrow [wait\_for($
$is\_time(t + 1\ min))]done(record(info\_incident(related\_info),\ log))))$
$\Rightarrow_{CRL} RD.994.99 : 21.1$

$CRL: R37.2$ $(done(incident(CARREL)) \wedge is\_time(t)$
$\wedge \neg done(record(info\_incident(related\_info),\ log))))$
$\rightarrow [wait\_for(is\_time(t + 1\ min))]CRL:V36 \Rightarrow_{CRL} RD.994.99 : 21.2$

The last set of rules is the result of introducing those rules coming from the context of the *e*-organization which do not refine previous rules but give new rules (CRL:R31.1 to CRL:R37.2). Note that, in the case of rules CRL:R34.1 to CRL:R37.2 we have also defined which is the maximum time delay that is acceptable to record any event (one minute).

Checking for Completeness

Once the rules for the security police have been defined, we should add some more rules in order to complete the model. As in the case of the organ and tissue allocation process, we should address two aspects:

- completing the precedence relations between actions: by introducing those relations that, because they are *common sense*, are not included in normatives. Figure 6.7 sketches the precedence relations defined by the current rules, and it identifies two missing relations:

  - any external entity has to get *access* to the system before trying to *input* or *query* data. We can express this as follows:

    $CRL: R38$ $[input(ext\_role,\ data)]access(ext\_role,\ CARREL)$
    $CRL: R39$ $[query(ext\_role,\ data)]access(ext\_role,\ CARREL)$

- refining the actions and predicates: introducing the domain knowledge in terms or rules that refine some of the predicates:

  - One example is to refine the predicate $avoid\_possible\_attack$ with rules about the different possible attacks and the security measures to be taken (most of them based in *encryption*).[12]

  - Another example is to refine the predicate $check\_access\_rights$ with a set of rules about resources, roles and different access levels. An option here is to add most the axioms and tautologies of the logic for Role-Based Access Control (RBAC) defined in [119].

---

[12]It is important to note here that, at this level, those rules only define which (encryption) measures to take for a given condition (e.g., *"to avoid that other entities read the data, use confidentiality measures"*; *"to avoid unauthorized entities to send data as if they were unauthorized user, use authentication methods"*) but not the final details of the technologies used (e.g. *"an asymmetric encryption with the sender encrypting the data with the private key"*; *"a symmetric encryption with the sender using the shared key with the recipient"*).

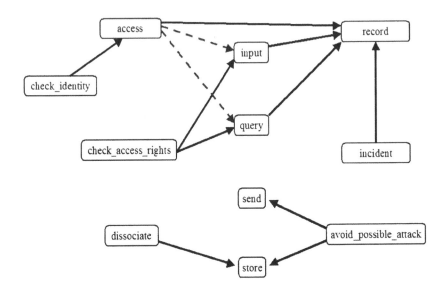

Figure 6.7: Sketch of the precedence relations among actions in the security allocation policy.

6.4.3 Violations and Sanctions

After refining all the rules that define which should be the behaviour of the system, now we focus on the violations that have been identified in previous sections, analyse them and define which are the sanctions.

To do so, first we will separate the violations coming from the behaviour of external entities (which we call *external violations*) from the ones related to the behaviour of the *CARREL* system (which we call *internal violations*). Then we will refine the conditions that define each violation, which are the possible sanctions and which are the additional actions to be performed to cope with the violation.

External Violations

In our framework, external violations are the ones where the designer should pay more attention. As we mentioned in Section 4.4, we cannot assume that agents entering into the *e*-organization will always follow the norms and rules imposed by its normative system. Therefore, an active enforcement should be made. In Section 4.4 we also proposed to create *police agents* in order to check such enforcement.

As, in our framework, a *police agent* does not have access to the internal beliefs, goals and intentions of the other agents, they can only check the agents' behaviour, by detecting when those agents enter in states considered *illegal* (see Section 4.2.1). The way of doing so is by means of the list of definitions of external violations. Such list defines, for each violation, the condition that triggers it. This condition is extracted from the rule that defines the violation. For instance, from rule CRL:R18.2:

$$CRL : R18.2 \quad (done(deliver(organ, hospital_1, hospital_2)) \wedge is\_time(t)$$
$$\wedge \neg done(hospital_2.send\_result\_transplant(recipient)))$$
$$\rightarrow [wait\_for(is\_time(t + 1 \ day))]CRL : V18.1$$

we can create the violation condition by stating that the action inside [ ] has been *done* [13]:

$$CRL : V18.1 \quad done(deliver(organ, hospital_1, hospital_2)) \wedge is\_time(t)$$
$$\wedge \neg done(hospital_2.send\_result\_transplant(recipient))$$
$$\wedge done(wait\_for(is\_time(t + 1 \ day)))$$

In the case of *CARREL*, such list is the following:

- Organ and tissue allocation policies:

$CRL:V14$ $\neg done(hospital.ensure\_quality(organ))$

$CRL:V15$ $\neg done(tissue\_bank.ensure\_quality(tissue))$

$CRL:V18.1$ $done(deliver(organ, hospital_1, hospital_2)) \wedge is\_time(t)$
     $\wedge done(wait\_for(is\_time(t + 1 \ day)))$
     $\wedge \neg done(hospital_2.send\_result\_transplant(recipient))$

$CRL:V18.2$ $done(deliver(organ, hospital_1, hospital_2)) \wedge is\_time(t)$
     $\wedge done(wait\_for(is\_time(t + 3 \ months)))$
     $\wedge \neg done(hospital_2.send\_evolution\_3months(recipient))$

$CRL:V18.3$ $done(deliver(organ, hospital_1, hospital_2)) \wedge is\_time(t)$
     $\wedge done(wait\_for(is\_time(t + 1 \ year)))$
     $\wedge \neg done(hospital_2.send\_evolution\_1year(recipient))$

$CRL:V18.4$ $done(deliver(tissue, tissue\_bank, hospital)) \wedge is\_time(t)$
     $\wedge done(wait\_for(is\_time(t + 1 \ day)))$
     $\wedge \neg done(hospital.send\_result\_transplant(recipient))$

$CRL:V18.5$ $done(deliver(tissue, tissue\_bank, hospital)) \wedge is\_time(t)$
     $\wedge done(wait\_for(is\_time(t + 3 \ months)))$
     $\wedge \neg done(hospital.send\_evolution\_3months(recipient))$

$CRL:V18.6$ $done(deliver(tissue, tissue\_bank, hospital)) \wedge is\_time(t)$
     $\wedge done(wait\_for(is\_time(t + 1 \ year)))$
     $\wedge \neg done(hospital.send\_evolution\_1year(recipient))$

---

[13]This translation can be performed, as the predicate *done* complies the following property: $[\alpha]done(\alpha)$ ("*after action α is performed, it is always the case that α is done*").

- Security policy:

$CRL:V24$ $\neg done(hospital.ensure\_accuracy(donor,\ data))$
$CRL:V25$ $\neg done(hospital.ensure\_accuracy(recipient,\ data))$
$CRL:V26$ $\neg done(tissue\_bank.ensure\_accuracy(donor,\ data))$
$CRL:V27$ $\neg done(hospital.avoid\_possible\_attack(donor\_data))$
             $\wedge done(hospital.send(donor\_data,\ CARREL))$
$CRL:V28$ $\neg done(hospital.avoid\_possible\_attack(recipient\_data))$
             $\wedge done(hospital.send(recipient\_data,\ CARREL))$
$CRL:V29$ $\neg done(tissue\_bank.avoid\_possible\_attack(donor\_data))$
             $\wedge done(tissue\_bank.send(donor\_data,\ CARREL))$

The above lists of violations define situations that are *illegal*. However, all of them (except CRL:V18.1 to CRL:V18.6) are described in terms of actions that cannot be checked directly by *CARREL* (e.g., that the hospital/tissue bank has ensured the quality of the organ, or has ensured the accuracy of the data). On the contrary, in the case of CRL:V18.1 to CRL:V18.6, all conditions can be checked by *CARREL*: the fact that an organ/tissue has been sent to a hospital in a given time, the fact that the time limit expired and the fact that such hospital has not sent the results of the recipient.

Therefore, those conditions that cannot be checked should be re-written in terms of actions that can be directly checked by *CARREL*. The result is the following:

$CRL:V14$ $\neg done(hospital.send\_quality\_certification(organ))$
$CRL:V15$ $\neg done(tissue\_bank.send\_quality\_certification(tissue))$
$CRL:V24$ $done(hospital.send(donor\_data,\ CARREL)) \wedge \neg accurate(donor\_data)$
$CRL:V25$ $done(hospital.send(recipient\_data,\ CARREL)) \wedge \neg accurate(recipient\_data)$
$CRL:V26$ $done(tissue\_bank.send(donor\_data,\ CARREL)) \wedge \neg accurate(donor\_data)$
$CRL:V27$ $done(hospital.send(donor\_data,\ CARREL)) \wedge \neg protected(donor\_data)$
$CRL:V28$ $done(hospital.send(donor\_data,\ CARREL)) \wedge \neg protected(recipient\_data)$
$CRL:V29$ $done(tissue\_bank.send(donor\_data,\ CARREL)) \wedge \neg protected(donor\_data)$

In CRL:V14 and CRL:V15 we have translated the unverifiable conditions of hospitals and tissue banks ensuring quality of pieces by one which is verifiable: that hospitals and tissue banks send some kind of certification that the piece has the quality needed. In violations CRL:V24 to CRL:V29 the scheme used is always the same: once the hospital/tissue bank sends the data to *CARREL*, the e-*organization* checks if the data is *accurate* (by checking if important values are missing or out of range and if the values are consistent one to another) or *protected* (e.g., encrypted) to avoid attacks.

Once the external violations have been identified and its pre-conditions defined, the next step is the definition of the *sanction* for each violation.

$CRL:S14$ $\{request(hospital,\ send\_quality\_certification(organ));$
            $inform(board,\ \neg done(hospital.send\_quality\_certification(organ))))\}$
$CRL:S15$ $\{request(tissue\_bank,\ send\_quality\_certification(tissue));$
            $inform(board,\ \neg done(tissue\_bank.send\_quality\_certification(tissue))))\}$

$CRL:S18.1$ $\{request(hospital_2, send\_result\_transplant(recipient));$
$inform(board, done(deliver(organ, hospital_1, hospital_2)) \land is\_time(t)$
$\land done(wait\_for(is\_time(t + 1\ day)))$
$\land \neg done(hospital_2.send\_result\_transplant(recipient))))\}$

$CRL:S18.2$ $\{request(hospital_2, send\_evolution\_3months(recipient));$
$inform(board, done(deliver(organ, hospital_1, hospital_2)) \land is\_time(t)$
$\land done(wait\_for(is\_time(t + 3\ months)))$
$\land \neg done(hospital_2.send\_evolution\_3months(recipient))))\}$

$CRL:S18.3$ $\{request(hospital_2, send\_evolution\_1year(recipient));$
$inform(board, done(deliver(organ, hospital_1, hospital_2)) \land is\_time(t)$
$\land done(wait\_for(is\_time(t + 1\ year)))$
$\land \neg done(hospital_2.send\_evolution\_1year(recipient))))\}$

$CRL:S18.4$ $\{request(hospital, send\_result\_transplant(recipient));$
$inform(board, done(deliver(tissue, tissue\_bank, hospital)) \land is\_time(t)$
$\land done(wait\_for(is\_time(t + 1\ day)))$
$\land \neg done(hospital.send\_result\_transplant(recipient))))\}$

$CRL:S18.5$ $\{request(hospital, send\_evolution\_3months(recipient));$
$inform(board, done(deliver(tissue, tissue\_bank, hospital)) \land is\_time(t)$
$\land done(wait\_for(is\_time(t + 3\ months)))$
$\land \neg done(hospital.send\_evolution\_3months(recipient))))\}$

$CRL:S18.6$ $\{request(hospital, send\_evolution\_1year(recipient));$
$inform(board, done(deliver(tissue, tissue\_bank, hospital)) \land is\_time(t)$
$\land done(wait\_for(is\_time(t + 1\ year)))$
$\land \neg done(hospital.send\_evolution\_1year(recipient))))\}$

$CRL:S24$ $\{inform(hospital, \neg accurate(donor\_data));$
$request(hospital, send(donor\_data));$
$inform(board, \neg accurate(donor\_data))\}$

$CRL:S25$ $\{inform(hospital, \neg accurate(recipient\_data));$
$request(hospital, send(recipient\_data));$
$inform(board, \neg accurate(recipient\_data))\}$

$CRL:S26$ $\{inform(tissue\_bank, \neg accurate(donor\_data));$
$request(tissue\_bank, send(donor\_data));$
$inform(board, \neg accurate(donor\_data))\}$

$CRL:S27$ $\{inform(hospital, \neg protected(donor\_data));$
$inform(board, \neg protected(donor\_data))\}$

$CRL:S28$ $\{inform(hospital, \neg protected(recipient\_data));$
$inform(board, \neg protected(recipient\_data))\}$

$CRL:S29$ $\{inform(tissue\_bank, \neg protected(donor\_data));$
$inform(board, \neg protected(donor\_data))\}$

In the domain of organ and tissue transplantation, it is non-sense to define a sanction system based on monetary fines (i.e., the hospital paying each time it enters into a violation) or negative scoring (i.e., the hospital scoring negative points when entering into a violation). In our case the sanction is implicit in the action of informing the ONT's board about the violation. This action has two effects: 1) the ONT's board may reduce its confidence in the hospital/tissue bank, if it enters in too many violations, 2) eventually the board might issue a corrective measure to the hospital/tissue bank to avoid further violations (e.g., imposing a deadline to a hospital to arrange its hardware/software to send all information encrypted and, in case the deadline expires, then the hospital may even temporary loose its permit to perform transplants). We also include, as part of the sanction, the action or actions to be carried out by the hospital/tissue bank.

To end the definition of the violation, we must also describe the actions to be triggered by the *e*-organization when the violation occurs. We will call such actions *side effects*:

$CRL:E14$ $\{stop\_assignation(organ);$
            $record(done(\neg hospital.send\_quality\_certification(organ)),incident\_log)\}$

$CRL:E15$ $\{stop\_assignation(tissue);$
            $record(done(\neg tissue\_bank.send\_quality\_certification(tissue)),incident\_log)\}$

$CRL:E18.1$ $\{record(done(deliver(organ, hospital_1, hospital_2)) \wedge is\_time(t)$
            $\wedge done(wait\_for(is\_time(t + 1\ day)))$
            $\wedge \neg done(hospital_2.send\_result\_transplant(recipient)),incident\_log)\}$

$CRL:E18.2$ $\{record(done(deliver(organ, hospital_1, hospital_2)) \wedge is\_time(t)$
            $\wedge done(wait\_for(is\_time(t + 3\ months))$
            $\wedge \neg done(hospital_2.send\_evolution\_3months(recipient)),incident\_log)\}$

$CRL:E18.3$ $\{record(done(deliver(organ, hospital_1, hospital_2)) \wedge is\_time(t)$
            $\wedge done(wait\_for(is\_time(t + 1\ year)))$
            $\wedge \neg done(hospital_2.send\_evolution\_1year(recipient)),incident\_log)\}$

$CRL:E18.4$ $\{record(done(deliver(tissue, tissue\_bank, hospital)) \wedge is\_time(t)$
            $\wedge done(wait\_for(is\_time(t + 1\ day)))$
            $\wedge \neg done(hospital.send\_result\_transplant(recipient)),incident\_log)\}$

$CRL:E18.5$ $\{record(done(deliver(tissue, tissue\_bank, hospital)) \wedge is\_time(t)$
            $\wedge done(wait\_for(is\_time(t + 3\ months)))$
            $\wedge \neg done(hospital.send\_evolution\_3months(recipient)),incident\_log)\}$

$CRL:E18.6$ $\{record(done(deliver(tissue, tissue\_bank, hospital)) \wedge is\_time(t)$
            $\wedge done(wait\_for(is\_time(t + 1\ year)))$
            $\wedge \neg done(hospital.send\_evolution\_1year(recipient)),incident\_log)\}$

$CRL:E24$ $\{record(\neg accurate(donor\_data),incident\_log); wait\_for(donor\_data)\}$

$CRL:E25$ $\{record(\neg accurate(recipient\_data),incident\_log); wait\_for(recipient\_data)\}$

$CRL:E26$ $\{record(\neg accurate(donor\_data),incident\_log); wait\_for(donor\_data)\}$

$CRL:E27$ $\{record(\neg protected(donor\_data),incident\_log)\}$

$CRL:E28$ $\{record(\neg protected(recipient\_data),incident\_log)\}$

$CRL:E29$ $\{record(\neg protected(donor\_data),incident\_log)\}$

As we can see in the definitions above, in all cases *CARREL* has to record the violation as an incident in its incident log. Then other actions might also be needed. For instance, in CRL:E14 and CRL:E15, the system must stop the assignation of the organ, as the quality certificate has not been received. There are other cases where no additional actions need to be carried out. For instance, in CRL:E27 to CRL:E29, the assignation process may continue although the received data was not properly protected by the sender.[14]

At this point we have all the components needed to define a violation at the Rule Level. Following there is an example of the complete definition of the CRL:V14 violation:

Violation: $CRL:V14$
Pre-conditions: ¬*done*(*hospital.send_quality_certification*(*organ*))
Sanction: { *request*(*hospital, send_quality_certification*(*organ*));
                inform(*board,* ¬*done*(*hospital.send_quality_certification*(*organ*)))}
Side-effects: { *stop_assignation*(*organ*);
                record(¬*done*(*hospital.send_quality_certification*(*organ*)),*incident_log*)}

This definition will be extended in the Procedure Level (see Section 6.5.3) to identify the role or roles inside *CARREL* which are responsible to check the conditions that define each violation.

Internal Violations

Once the external violations are identified and properly defined, the next step is to proceed with the internal violations. Internal violations describe states that the *e*-organization should always avoid. As the designer has full control of the design of the agents inside the *e*-organization, in this case the agents, as fully agree with the objectives of the *e*-organization, will follow its norms and rules. So, in this case the creation of the *police agent* role and the definition of the violations does not aims to create an *enforcement mechanism* but a continuous *safety control* of the system's behaviour (i.e., avoid the system to enter in a undesirable, *illegal* state because of a failure in one of the agents).

The list of identified internal violations is the following:

- Organ and tissue allocation policies:

$CRL:V1$ ¬*done*(*get_quality*(*organ, hospital*)) ∧ *done*(*assign*(*organ, recipient*))
$CRL:V2$ ¬*done*(*get_quality*(*tissue, tissue_bank*)) ∧ *done*(*assign*(*tissue, recipient*))
$CRL:V3$ ¬*done*(*ensure_compatibility*(*organ, recipient*))
         ∧ *done*(*assign*(*organ, recipient*))
$CRL:V4$ ¬*done*(*ensure_compatibility*(*tissue, recipient*))
         ∧ *done*(*assign*(*tissue, recipient*))
$CRL:V5$ ¬*done*(*seek_time_efficiency*(*distribution*))
$CRL:V6$ *done*(*assign*(*organ, recipient*)) ∧ *done*(*use*(*recipient, race*))
$CRL:V7$ *done*(*assign*(*tissue, recipient*)) ∧ *done*(*use*(*recipient, race*))
$CRL:V8$ *done*(*assign*(*organ, recipient*)) ∧ *done*(*use*(*recipient, sex*))
$CRL:V9$ *done*(*assign*(*tissue, recipient*)) ∧ *done*(*use*(*recipient, sex*))

---

[14]This solution is dangerous, as it might happen that the unprotected data was changed during transmission by a unauthorized third party. The ideal solution would be to ask to all hospitals and tissue banks to use encryption mechanisms when sending data through a network, but in practice hospitals and tissue banks would need an adaptation time to meet this requirement.

$CRL:V10$ $\neg done(plan\_delivery(organ,\ hospital_1,\ hospital_2))$
          $\wedge\ done(deliver(organ,\ hospital_1,\ hospital_2))$

$CRL:V11$ $\neg done(plan\_delivery(tissue,\ tissue\_bank,\ hospital))$
          $\wedge\ done(\wedge deliver(tissue,\ tissue\_bank,\ hospital))$

$CRL:V12$ $\neg done(inform(board,\ distribution))$

$CRL:V13$ $\neg done(inform(board,\ logistics))$

$CRL:V16$ $\neg done(check\_if\_urgency\_0(organ)) \wedge done(assign(organ,\ patient))$

$CRL:V17$ $\neg done(check\_if\_urgency\_0(tissue)) \wedge done(assign(tissue,\ patient))$

- Security policy:

$CRL:V19.1$ $\neg done(dissociate(donor\_id,\ data)) \wedge done(store(data,\ memory))$

$CRL:V19.2$ $\neg done(dissociate(donor\_id,\ data)) \wedge done(store(data,\ disc))$

$CRL:V20.1$ $\neg done(dissociate(recipient\_id,\ data)) \wedge done(store(data,\ memory))$

$CRL:V20.2$ $\neg done(dissociate(recipient\_id,\ data)) \wedge done(store(data,\ disc))$

$CRL:V21.1$ $\neg done(avoid\_possible\_attack(recipient\_waiting\_lists))$
          $\wedge\ done(store(recipient\_waiting\_lists,\ disc))$

$CRL:V21.2$ $\neg done(avoid\_possible\_attack(recipient\_waiting\_lists))$
          $\wedge\ done(store(recipient\_waiting\_lists,\ memory))$

$CRL:V21.3$ $\neg done(avoid\_possible\_attack(recipient\_waiting\_lists))$
          $\wedge\ done(send(recipient\_waiting\_lists,\ network))$

$CRL:V22.1$ $\neg done(avoid\_possible\_attack(donor\_data))$
          $\wedge done(store(donor\_data,\ disc))$

$CRL:V22.2$ $\neg done(avoid\_possible\_attack(donor\_data))$
          $\wedge\ done(store(donor\_data,\ memory))$

$CRL:V22.3$ $\neg done(avoid\_possible\_attack(donor\_data))$
          $\wedge\ done(send(donor\_data,\ network))$

$CRL:V23.1$ $\neg done(avoid\_possible\_attack(recipient\_data))$
          $\wedge\ done(store(recipient\_data,\ disc))$

$CRL:V23.2$ $\neg done(avoid\_possible\_attack(recipient\_data))$
          $\wedge\ done(store(recipient\_data,\ memory))$

$CRL:V23.3$ $\neg done(avoid\_possible\_attack(recipient\_data))$
          $\wedge\ done(send(recipient\_data,\ network))$

$CRL:V30$ $\neg done(check\_identity(agent) \wedge done(access(agent,\ CARREL))$

$CRL:V31$ $\neg done(check\_access\_rights(agent,\ data)) \wedge done(input(agent,\ data)$

$CRL:V32$ $\neg done(check\_access\_rights(agent,\ data)) \wedge done(query(agent,\ data))$

$CRL:V33$ $done(access(agent,\ CARREL)) \wedge is\_time(t)$
          $\wedge\ done(wait\_for(is\_time(t + 1\ min)))$
          $\wedge \neg done(record(info\_access(agent,\ related\_info),\ log)))$

$CRL:V34$ $done(input(agent,\ data)) \wedge is\_time(t)$
          $\wedge done(wait\_for(is\_time(t + 1\ min)))$
          $\wedge \neg done(record(info\_input(agent,\ related\_info),\ log)))$

$CRL:V35$ $done(query(agent,\ data)) \wedge is\_time(t)$
          $\wedge\ done(wait\_for(is\_time(t + 1\ min)))$
          $\wedge \neg done(record(info\_query(agent,\ related\_info),\ log)))$

$CRL:V36$ $done(incident(CARREL)) \wedge is\_time(t)$
         $\wedge \; done(wait\_for(is\_time(t + 1 \; min)))$
         $\wedge \neg done(record(info\_incident(related\_info), log)))$

The violations in the lists above can be classified in three groups:

- *violations of precedence between actions or states*: these are violations coming from precedence rules. As the agents inside *CARREL* will include those rules in their reasoning process, the related violations will, in principle, only occur when there is a failure in one of the agents (e.g., the connection with the agent suddenly breaks). The violations in this group are the following: CRL:V1 to CRL:V4, CRL:V10 to CRL:V11, CRL:V16, CRL:V17, CRL:V19.1 to CRL:V23.3 and CRL:V30 to CRL:V36. The structure of the complete violation is always the same:

```
Violation:        CRL:V19.1
Pre-conditions:   ¬done(ensure_compatibility(organ, recipient))
                  ∧done(assign(organ, recipient))
Sanction:         {inform(board, ¬done(ensure_compatibility(organ, recipient))
                  ∧done(assign(organ, recipient)))}
Side-effects:     {stop_assignation(organ);
                  record(¬done(ensure_compatibility(organ, recipient))
                  ∧done(assign(organ, recipient)),
                  incident_log);
                  wait_for(done(ensure_compatibility(organ, recipient)))}
```

The precondition is the set of conditions that trigger the violation. Usually, they are already refined, so no more refinement is needed at this level. Then the sanction consists only in informing the board about the incident, as it has no sense to define some kind of punishment for the agents inside *CARREL*. Finally, the side-effects are the set of actions to solve the situation (that is, a contingency plan). It has usually the same structure: stop the action related to the second predicate, record the incident in the incident log and then wait to the first action to be performed.

- *violations that are abstract*: these are violations defined by quite abstract conditions which have not been refined in the previous steps, as they refer to abstract concepts or concepts that are hard to translate in one or several actions. In our case we have three of those violations:

$CRL:V5$ $\neg done(seek\_time\_efficiency(distribution))$
$CRL:V12$ $\neg done(inform(board, distribution))$
$CRL:V13$ $\neg done(inform(board, logistics))$

In the case of CRL:V5, the condition is hard to translate to a set of actions, as it is quite abstract (i.e., *when can we say that the system stopped seeking for time efficiency?*). In fact it declares a quite abstract requirement for the implementation step: that time efficiency should be kept whenever possible. In the case of CRL:V12 and

CRL:V13, they state that the board should be informed about the (important) events of the distribution and the logistics. This is again a quite abstract statement, useful as a requirement for the implementation step to be met by the designer: to design the system in a way that members of ONT's board have some kind of interface to check the system performance and also be proactively informed of any (important) event or incident.

- *violations enforcing specific requirements*: these are violations defined by specific conditions which should be met by the designer. In our example these violations are:

$$CRL:V6 \quad done(assign(organ, recipient)) \land done(use(recipient, race))$$
$$CRL:V7 \quad done(assign(tissue, recipient)) \land done(use(recipient, race))$$
$$CRL:V8 \quad done(assign(organ, recipient)) \land done(use(recipient, sex))$$
$$CRL:V9 \quad done(assign(tissue, recipient)) \land done(use(recipient, sex))$$

In this case, the conditions say that it is a violation to use information about race or sex during the assignment. Therefore, this is a requirement to be met, for instance, by avoiding the agents taking part in the assignation process to receive information about race or sex.

6.4.4   Summary: the Rule Level in *CARREL*

Summarizing the results of the previous sections, the rule level of *CARREL* can be described as follows:

$$rulelevel = \langle rules, violations, ontology_{rl} \rangle$$

where:

- $rules = \langle CRL:R1.1, CRL:R1.2, \ldots CRL:R39, \rangle \cup domain\_rules$,

- *domain_rules* are the rules added to model de domain,

- $violations = \langle CRL:V1 \ldots CRL:V4, CRL:V10, CRL:V11,$
  $CRL:V14 \ldots CRL:V36 \rangle$

- $ontology_{rl}$ is the ontology that defines the terms and predicates appearing in the *rules* set.

6.5   The Procedure Level in *CARREL*

Once the Rule Level is defined, the next step is to implement a system (triggers, procedures, protocols) which follows the rules and checks the violations that have been identified.

First, we will present which are the implementation decisions taken, including those related to some abstract requirements identified in the Rule Level. Then we will refine the role hierarchy and the violations to adapt them to the decisions taken. Finally, we will sketch how we implement the system by using the ISLANDER formalism.[15]

6.5.1 Implementation Decisions

At this point there are some generic decisions to be taken about how we will implement the system at the Procedure Level:

- *I1: Implement CARREL as a multi-agent system*: this first decision may seem obvious, because our framework's main aim is to build agent-mediated *e*-organizations where the agents follow a given normative system. But all the previous analysis made for *CARREL* is also valid for a standard software solution for *CARREL*: all the analysis in the previous levels refers to *CARREL* as a whole, and defines the responsibilities and the violations to be checked. Therefore an alternative implementation of *CARREL* could be an application which executes the rules in the previous levels. The choice to implement *CARREL* as a system composed by (FIPA-compliant) agents has been taken in order to build a distributed solution which uses, in its full potential, the defined rules and violations to coordinate the allocation process.

- *I2:Use a database to store all the data*: as the *CARREL* system has to manage huge amounts of data in a secure way and it also has to allow concurrent access to the data, we have chosen a database package such as Oracle in order to take profit of the security measures it embeds.

- *I3: Rule-driven vs. Procedure-driven*: another important decision to be taken is if the agents will directly behave by interpreting the rules or if they will follow a protocol.

  - The *rule-driven* agents option is the best one in an ideal situation where all the incoming agents are able to interpret the rules and the consequences of violations and where the agents have the resources needed to do the reasoning. But this assumption on the incoming agents' internal architecture and reasoning capabilities is too strong;

  - The *procedure-driven* agents option is more suitable for real applications where agents should behave and take decisions in real-time. But in this scenario the autonomy of the agents almost disappears, as they are not allowed to break the protocol at anytime. And as we saw in Section 5.5, the system may fail when enters in situations that the protocol did not foresee.

  Our choice is a mixed *rule-procedure-driven* approach, where *Flexible Normative Agents* (see Section 5.5) mostly follow the defined protocols, but they can eventually check the rules and decide to break the protocol in a way that a) is *legal* as it breaks no rule b) is *illegal* but the consequences of the violation are accepted. This feature is very valuable, for instance, in situations where one step of the protocol is delayed because of some problem (e.g., a broken connection to the provider of some

---

[15] We also have some preliminary work about the use of PRO*forma*/Cogent to implement *CARREL*'s tissue allocation policy [20].

information) and the rules allow to break the protocol and pass to the next steps as far as the precedence restrictions in the rules are met.

As part of the design process, we have to check also some of the requirements that we identified in Section 6.4.3. They are summarized in the following list:

$CRL:V5$ $\neg done(seek\_time\_efficiency(distribution))$
$CRL:V12$ $\neg done(inform(board, distribution))$
$CRL:V13$ $\neg done(inform(board, logistics))$
$CRL:V6$ $done(assign(organ, recipient)) \wedge done(use(recipient, race))$
$CRL:V7$ $done(assign(tissue, recipient)) \wedge done(use(recipient, race))$
$CRL:V8$ $done(assign(organ, recipient)) \wedge done(use(recipient, sex))$
$CRL:V9$ $done(assign(tissue, recipient)) \wedge done(use(recipient, sex))$

These requirements are the origin of additional decisions:

- CRL:V5 states a requirement about designing the system in a way that is as time efficient as possible. This requirement results in two decisions:

  - *I4: Parallel processes*: We will try to run processes as parallel as possible. In the case of organ allocation we will split the *assign* process in two parts:
    1. a *pre-assignment step*, where a search for compatible recipients of a given organ or tissue is performed, even if the donation is not confirmed yet (e.g., quality of the organ has not been received).[16] This also allows to start building the delivery plan as soon as possible.

    2. an *assignation step*, once all the conditions to meet (delivery plan built, no urgency-0) are fulfilled.

  - *I5: Time control*: We will create clocks to control maximum times for each step and for the full process, so an alarm may rise if, e.g.:
    * an organ offer has not been assigned and is reaching the time limit recommended for such organ to be out of the human body;

    * a hospital has not sent the quality certification of the organ;

- *I6: Interface*: From CRL:V12 and CRL:V13, we can extract the requirement of creating a proper interface between the system and the board members. Such interface should allow the board to check the state of all the transactions and also allow *CARREL* to notify the board when an incident or an important event appears. In order

---

[16]This is a process that also is done in practice in the Spanish Health system: in order to reduce the time from the extraction to the implant of an organ, the hospital where there is a possible donor sends a provisional offer of organs to the ONT or the OCATT. Then, some times the search process already starts and the hospital which would receive the organ is informed about the possible donation, in order to give that hospital the time to organize the transplant (call the *candidate* recipient to come to the hospital and do some tests while the offer is confirmed). Then it may happen that the offer is confirmed (so everything and everyone is in place to take part in the process in the destination hospital, and the extraction-transport-implant process is optimized in time) or denied (because of the quality tests to the organ once it has been extracted, or because of a denial of the family to donate the organs).

to do so, the interface between *CARREL* and the board may use one or several of the following technologies: a standard computer application, messages to beepers, SMS messages or WAP interfaces to mobile phones, WAP or wireless connections to PDA's.

- *I7: No discrimination*: In CRL:V6 to CRL:V9 there is a requirement that the agents involved in the pre-assignation and assignation will never use data about sex or race of recipients (unless further developments in medical knowledge may demonstrate that such factors are important for a specific kind of piece and a given implantation technique). In order to meet this requirement, we will avoid that such data, if present, is accessible from the agents that are involved in the assignation.

6.5.2 Refining the Role Hierarchy

Once the general decisions about implementation have been made, we have to extend the role hierarchy to introduce the impact of those decisions. In our case, as we have decided to implement *CARREL* as a multi-agent system, this means to identify the roles to be enacted by software agents and the relations between those roles.

We will only do the analysis for those goals where the *CARREL* role is involved (*distribute tissues* and *distribute organs*).

In the case of the *distribute tissues* goal, we will use as starting point the goal distribution already defined in figure 6.3. The first step is to refine the *CARREL* role. Such role has, as assigned goal, to manage tissue requests and offers. Such goal is then distributed in the following roles:

- *admission*: to authenticate the agents coming from hospitals and tissue banks to offer or request tissues. Therefore, is the role related with access control to the system,

- *allocator*: to assign organs and tissues, ensuring compatibility between donor and recipient,

- *planner*: to build the delivery plan (from a hospital or a tissue bank to another hospital).

- *data manager*: to ensure (role based) access control to the data stored in the database.

- *police*: to detect when the behaviour of the agents inside *CARREL* may create a violation state and then act properly (i.e., execute the defined *sanctions* and *side-effects* for that violation).

It is important to note that all the new roles have a *part-of* relation with the *CARREL* role (the sub-roles *compose* the whole *CARREL* system).

We also have to refine the other roles in the role hierarchy involved in the *distribute tissues* goal. In the case of the *tissue bank*, as it only has to send information about the available tissues, the goal can be given to a single agent role: *tissue bank coordinator*.

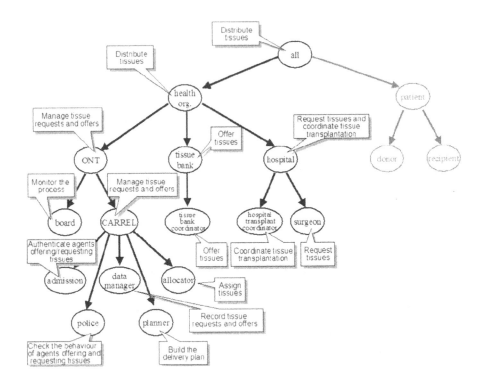

Figure 6.8: Extension of the role hierarchy for the *distribute tissues* goal.

In the case of the *hospitals*, the assigned goal (to request tissues and coordinate tissue transplantation) is split into the *request tissues* goal (*surgeon* role) and the *coordinate tissue transplantation* goal (*hospital transplant coordination* role). Again, these new sub-roles identified for the tissue banks and the hospitals have a *part-of* relation with their super-role.

The case of the *distribute organs* goal distribution is very similar to the one described for tissues and it is shown in figure 6.9.

As a result of the refinement in the goal distribution, the new roles, which are the ones to be enacted by software agents,[17] are the following:

$$agent\_roles = \quad \langle police, admission, allocator, data\_manager,$$
$$planner, tissue\_bank\_coordinator,$$
$$hospital\_transplant\_coordinator \rangle$$

---

[17]One might think that, once the agent roles are identified, the rest of the roles in the role hierarchy have no value at the Procedure Level. But as we will see in Section 6.6, these roles allow us to express conditions that concern to those roles, independently of the subroles that may represent them inside the *e*-organization.

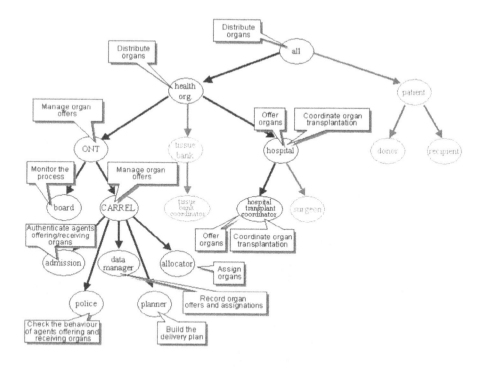

Figure 6.9: Extension of the role hierarchy for the *distribute organs* goal.

Finally, we have also to identify the new power relations between the new roles. As in the case of the subroles of *CARREL* there is again a *part-of* relation, we have no automatic inheritance of the power relations, so we have to determine them depending on the capabilities of the subroles. In this case:

- All the roles that might interact with external roles (*police. admission, allocator* and *data manager*) should be empowered over the external roles.

- We should empower the *police* role over the rest of roles in *CARREL*.

These power relations extend the list we identified in Section 6.2 and give as result the following one:

$$power\_rel = \quad \langle \langle ONT, hospital \rangle , \langle ONT, tissue\_bank \rangle , \langle board, hospital \rangle ,$$
$$\langle board, tissue\_bank \rangle , \langle CARREL, hospital \rangle ,$$
$$\langle CARREL, tissue\_bank \rangle , \langle police, admission \rangle ,$$
$$\langle police, allocator \rangle , \langle police, data\_manager \rangle , \langle police, planner \rangle ,$$
$$\langle police, tissue\_bank\_coordinator \rangle ,$$
$$\langle police, hospital\_transplant\_coordinator \rangle ,$$
$$\langle admission, tissue\_bank\_coordinator \rangle ,$$
$$\langle admission, hospital\_transplant\_coordinator \rangle ,$$
$$\langle allocator, tissue\_bank\_coordinator \rangle ,$$
$$\langle allocator, hospital\_transplant\_coordinator \rangle ,$$
$$\langle data\_manager, tissue\_bank\_coordinator \rangle ,$$
$$\langle data\_manager, hospital\_transplant\_coordinator \rangle \rangle$$

6.5.3   Refining the Violations and Sanctions

Once the role hierarchy has been extended with the roles to be enacted by the agents, the definition of the violations can be completed by identifying the role or roles inside the *CARREL* system that have to check the violation.

As seen in Section 6.2, there is a role in the role hierarchy, the *police* role, which main goal is to ensure the proper behaviour of all the agents in the *CARREL* system. Therefore, the agent or agents enacting such role are the ones responsible for the control of violations. however, through the power relation among roles, part of the violation control can be delegated from the *police* role to other roles which have the needed capabilities to check a given violation. For instance, to check CRL:V24 (hospitals sending non-accurate data about donors), all agents enacting roles which might receive such data (*admission role, allocator role*) should be able to perform this check directly (or submit it to another agent to be checked).

The resulting list of duties, describing for each violation the roles which should check it, is the following:

- external violations:

$CRL:V14$	*police, allocator*
$CRL:V15$	*police, allocator*
$CRL:V18.1$	*police*
$CRL:V18.2$	*police*
$CRL:V18.3$	*police*
$CRL:V18.4$	*police*
$CRL:V18.5$	*police*
$CRL:V18.6$	*police*
$CRL:V24$	*police, admission, allocator*
$CRL:V25$	*police, admission, allocator*
$CRL:V26$	*police, admission, allocator*

$CRL:V27$ *police, admission, allocator*
$CRL:V28$ *police, admission, allocator*
$CRL:V29$ *police, admission, allocator*

- internal violations:

$CRL:V1$	*police, allocator*	$CRL:V30$	*police, admission*
$CRL:V2$	*police, allocator*	$CRL:V31$	*data_manager*
$CRL:V3$	*police, allocator*	$CRL:V32$	*data_manager*
$CRL:V4$	*police, allocator*	$CRL:V33$	*police, admission*
$CRL:V10$	*police, allocator, planner*	$CRL:V34$	*data_manager*
$CRL:V11$	*police, allocator, planner*	$CRL:V35$	*data_manager*
$CRL:V16$	*police, allocator*	$CRL:V36$	*police*
$CRL:V17$	*police, allocator*		
$CRL:V19.1$	*data_manager*		
$CRL:V19.2$	*data_manager*		
$CRL:V20.1$	*data_manager*		
$CRL:V20.2$	*data_manager*		
$CRL:V21.1$	(the database's security system)		
$CRL:V21.2$	(the security of the agent platform)		
$CRL:V21.3$	*data_manager, allocator*		
$CRL:V22.1$	(the database's security system)		
$CRL:V22.2$	(the security of the agent platform)		
$CRL:V22.3$	*data_manager, allocator*		
$CRL:V23.1$	(the database's security system)		
$CRL:V23.2$	(the security of the agent platform)		
$CRL:V23.3$	*data_manager, allocator*		

In the case of the internal violations, we can see that we have dropped out the *police* role from some violation checks (the ones about data in disc and data sent though the network). This is done to avoid the *police agents* (that is, agents enacting the *police* role) to be intensively checking any query or input of data to the database or any message sent from the agents inside the system to the agents from outside (message passing is done through the network). To avoid such bottleneck, in these cases the roles that are directly involved in such activities are empowered with the full checking of these violations, allowing the *police agents* to focus on other conditions' checks that may affect more important processes (such as a missing step in the allocation process).

It is also important to note that, in our case, some of the violations are not checked by agents but by the operating system (memory access), the database (access to data in the database files) and the agent platform (memory access). But this is because the technologies we use (UNIX, ORACLE and Java-JADE), which provide the needed security measures. In the case of violations CRL:V21.3, CRL:V22.3 and CRL:V23.3 (all them concerning security in the message passing), this should be checked by the agents, as right now communications' security in security systems such as JADE-S is still under development.

At this point we have all the components needed to define a violation in our framework. Following there is an example of the complete definition of the CRL:V14 violation:

Violation: $CRL:V14$
Pre-conditions: $\neg done(hospital.send\_quality\_certification(organ))$
Sanction: $\{request(hospital,send\_quality\_certification(organ));$
                  $inform(board, \neg done(hospital.send\_quality\_certification(organ)))\}$
Side-effects: $\{stop\_assignation(organ);$
                  $record(\neg done(hospital.send\_quality\_certification(organ)),incident\_log)\}$
Enforcing roles: $\{police, allocator\}$

6.5.4   The Procedure Level in ISLANDER

In this section we will now define the protocols of the *CARREL* system through the IS-LANDER formalism. As we saw in Section 6.5.1, we have chosen a mixed *rule-protocol driven* approach where protocols are defined in order to ease the communication among agents and to speed up the process, but we will also keep the rules defined in the Rule Level to allow agents to, eventually, break the protocol to cope with unexpected situations.

Therefore, here we will use ISLANDER to specify such protocol and to implement the final MAS. This process is guided by all the specification of roles, norms and rules in the Concrete Level:

- The role hierarchy in the Concrete Level already identifies the roles that agents will play in ISLANDER, and guides the election of the agents that will enact each of these roles.

- The precedence rules in the Concrete Level can be used as a base to build the interaction protocols, by mapping interaction states with the predicates in the rules, and ensuring that precedence relations hold.

We already have a specification in ISLANDER of *CARREL*'s protocols (presented in Section 2.5) and an implementation of those protocols in an agent platform, though. So, instead of building a brand new specification and implementation from scratch, we will use the prototype that we already have as a starting point. This will also allow us to check that those protocols and the resulting implementation follow the regulations we have found in Section 6.4 (that is, that they follow the rules defined in the Rule Level) and to identify extensions to be done to the prototype we have implemented. [18]

Implementation Decisions

Some of the implementation decisions we have mentioned in Section 6.5.1 were already taken by CARREL 2.0:

- *I*1 (multi-agent system): CARREL 2.0 was implemented in a FIPA-compliant JADE agent platform (see Section 2.5.5).

---

[18] We will refer to the previous specification and implementation of *CARREL* as CARREL 2.0.

- *I2* (storage in database): we use ORACLE to store all the information about organs, tissues, donors, recipients and waiting lists.

- *I4* (pre-assignation and assignation): pre-assignation is performed in the *Tissue Exchange Room* and the *Organ Exchange Room* (for tissues and organs, respectively) while the final assignation is done in the *Confirmation Room* (see Section 2.5.1).

- *I6* (interface): CARREL 2.0 had a first interface prototype to be used in computers, but in the future we plan to extend its functionalities and the technologies used.

- *I7* (no discrimination): in CARREL 2.0 no information about recipients' race and sex was used, neither in the pre-assignation nor in the final assignation.

The rest of implementation decisions do not invalidate CARREL 2.0 but define extensions to be made:

- *I3* (rule-protocol driven approach): CARREL 2.0 was mainly protocol-driven. Now we aim to extend it in a way that internal agents and external agents may also use rules in their reasoning cycle.

  - Extension 1: add, inside the CARREL 2.0 platform, the data structures representing the rules and the violations, so a) internal agents can reason about them, and b) the external agents receive such rules and violation definitions once they enter into the system.

- *I5* (time control): we will add in the next version of *CARREL* clocks to detect situations where unexpected delays appear.

  - Extension 2: add inside the CARREL 2.0 platform the time control mechanisms, by means of timers to detect that a certain time limit has expired.

Apart from the implementation decisions mentioned above, there are two additional decisions that were taken:

- *I8: Checking for accuracy of inputs as soon as possible*: in order to avoid the situation where an external agent enters the system carrying incomplete and/or inconsistent data (which may, in fact, stop the assignation process), the external agents provide almost all their data in the reception room, which is then checked to ensure its accuracy (that is, the data is well-formed, complete and consistent). If the information provided by an agent is not accurate, then it has to leave the *e*-organization. This implementation decision covers rules CRL:R25.1, CRL:R25.2 and CRL:R26 about ensuring accuracy of the data provided by hospitals and tissue banks.

- *I9: Creating a time span to manage urgency-0 requests*: as there is a separation of the assignation process in a provisional assignation (done in the *Organ* an *Tissue Exchange Rooms*) and a definitive one (done in the *Confirmation Room*), the time needed to create a delivery plan is used also to wait for any urgency-0 request that may need the organ or tissue that is provisionally assigned.

Correspondence among CARREL 2.0 Roles and *CARREL* **Roles**

Not surprisingly, the roles we identified in the definition of CARREL 2.0 are very similar to the ones we have obtained by the HARMON*IA* refinement process:

- internal roles[19] (roles for agents that carry out the management of the *e*-organization):

  - *Reception Room Manager* (rrm): manager of the admission of external agents, it also checks if requests are well-formed. It corresponds to the *admission* role in *CARREL*.

  - *Consultation Room Manager* (crm): manager of the access of all agents to the database. It corresponds to the *data manager* in *CARREL*.

  - *Tissue Exchange Room Manager* (trm): it does the pre-assignment of tissues according to compatibility factors. It partially matches the *allocator* role in *CARREL*.

  - *Organ Exchange Room Manager* (orm): it does the pre-assignment of organs according to compatibility factors. It partially matches the *allocator* role in *CARREL*.

  - *Confirmation Room Manager* (cfrm): it does the final assignation of organs and tissues. It corresponds to the *allocator* role in *CARREL*, except that the compatibility assessment has been done by the *Tissue Exchange Room Manager* and the *Organ Exchange Room Manager*.

  - *Institution Manager* (im): a role played by a single agent that registers all the events that happen inside *CARREL* and eventually coordinates all the scene managers when the system is entering in a unsafe state. This role corresponds to the *police* role in *CARREL*.

  We can see, in fact, the *Tissue Exchange Room Manager*, *Organ Exchange Room Manager* and *Confirmation Room Manager* roles as subroles of the *allocator* role in *CARREL*.

- external roles (roles for incoming agents):

  - *Hospital Finder Agent* (hf): role of agents sent by hospitals with tissue requests or organ offers that are seen from the point of view of the institution as requests for finding an acceptable tissue or recipient, respectively. It partially matches the *hospital transplant coordinator* role in *CARREL* (see below).

  - *Hospital Contact Agent* (hc): agents from a certain hospital that are contacted by the institution when an organ has appeared for a recipient that is on the waiting list of that hospital. The agent then enters the institution to accept the organ and to receive the delivery plan. It partially matches the *hospital transplant coordinator* role in *CARREL* (see below).

---

[19]We used to call them *institutional roles*, before we stated our distinction between institutions and organizations.

– *Hospital Information Agent* (hi): agents sent by hospitals to keep the *CARREL* system updated about any event related to a piece or the state of the waiting lists. They can also perform queries on the *CARREL*'s database. It almost matches the *hospital transplant coordinator* role in *CARREL* (see below).

– *Tissue bank notifier* (tb): agents sent by tissue banks in order to update *CARREL* about tissue availability. It corresponds to the *tissue bank coordinator* role in *CARREL* .

In this case, the *Hospital Finder Agent, Hospital Contact Agent* and *Hospital Information Agent* can be seen as subroles of *CARREL*'s *hospital transplant coordinator* role.

Therefore, we can see that the roles defined in CARREL 2.0 not only match the roles identified in *CARREL*, but that CARREL 2.0 even refines some of those roles, as some goals were split in sub-goals to be fulfilled by different types of agents.

Checking the Protocols

We will now check if the protocols defined within CARREL 2.0's performative structure complies with the rules we identified for *CARREL*. To do so we should first identify the relation between the states defined in the protocols and the predicates in the rules, and then check if the precedence relations defined in *CARREL*'s rules hold in CARREL 2.0's protocols.

There are some *CARREL* predicates that are easy to relate with states in CARREL 2.0:

- check_identity: this predicate is related to the part of the protocol that does the identification of all the external agents. In CARREL 2.0, the identification of external agents is done in states a_0 and a_1 in the *Reception Room*'s conversation graph;

- access: this predicate is related with the access control where agents are allowed or denied to continue inside the *CARREL* system. In CARREL 2.0, this is done in states a_2, a_6, w_1 and w_2 in the *Reception Room*'s conversation graph;

- accurate: as we stated in decision *18*, this predicate is placed in the part of the protocol that checks if the information sent by external agents is accurate. Most of this task is performed in states a_3, a_4, a_7 and w_3 in the *Reception Room*'s conversation graph, as is in this point where external agents are distributed to one or other branch of the conversational graph depending on the roles they enact, or they are forced to exit the *CARREL* system in case that they were not identified as authorized agents (i.e., they had not the authorizations needed).

- dissociate and store: these predicates are related to security measures to be taken when data is stored in *CARREL*'s database. In CARREL 2.0 this only happens

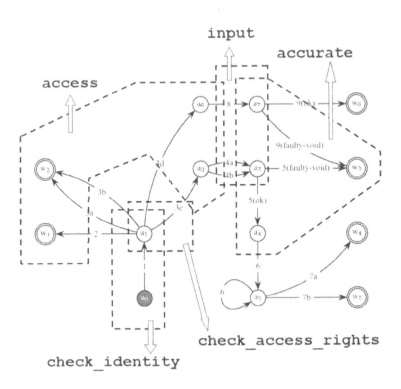

Figure 6.10: Some links from the predicates to the Reception Room.

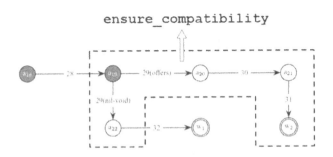

Figure 6.11: Some links from the predicates to the Tissue Exchange Room.

in the *Consultation Room* (state a_9), when the agent enacting the *Consultation Room Manager* role receives updates from agents coming from hospitals or tissue banks;

- query: this predicate is related with those states in the conversation graph where external agents are allowed to submit queries to *CARREL*'s database. In CARREL 2.0 this only happens in the *Consultation Room* (state $a_1 0$), when the agent enacting the *Consultation Room Manager* role receives queries from agents coming from hospitals or tissue banks;

- ensure_compatibility: this predicate is related to the part of the protocol that checks the different compatibility factors between recipients and pieces. In the case of tissues, this task is done inside the *Tissue Exchange Room* (states $a_1 8$, $a_1 9$, $a_2 0$, $a_2 1$, $a_2 2$, w_1 and w_2 of the conversation graph), while, in the case of organs, such task is done inside the *Organ Exchange Room* (states $a_1 2$, $a_1 4$, $a_1 7$ and w_3 of the conversation graph);

- deliver: this predicate is located in the part of the protocol where the *CARREL* system notifies to a hospital that a piece has been assigned to the patient, and also gives the consent to start the transportation from the source organization (a tissue bank or another hospital) to the destination. In CARREL 2.0, this situation happens in the *Confirmation Room*, state w_2;

- plan_delivery: this predicate corresponds to those states where the delivery plan is being built. In CARREL 2.0's conversation graphs, this occurs inside the *Confirmation Room*, from the moment the organ or tissue has been formally pre-assigned (state $a_2 3$) to the moment the assignation is confirmed (state $a_2 4$) or an urgency_0 request invalidated the pre-assignation and the delivery plan for that recipient and piece is no longer needed (state $a_2 5$);

- check_if_urgency_0: this predicate is also located in the time span between pre-assignation and assignation, so we can identify this predicate with states $a_2 3$ and $a_2 5$ in the *Confirmation Room*'s conversation graph.

There are other predicates that are harder to place, as 1) they are distributed in several parts of the protocol, and 2) they usually occur inside a single state of the protocol, as they refer to some actions to be performed by the agents as part of their inner reasoning cycle. In our case those predicates are the following:

- avoid_possible_attack and send: both predicates are be placed in any state where agents send important data (about patients or donors) to external agents. There are several states where the previous condition holds, but the main ones are: $a_1 0$ (*Consultation Room*), $a_1 3$ (*Organ Exchange Room*) and $a_2 4$ (*Confirmation Room*);

- input: in this case this predicate is placed in any state where agents receive as input important data about donors and recipients. There are several states, but the main ones are: a_3 and a_7 (*Reception Room*), a_9 (*Consultation Room*) and $a_1 5$ (*Organ Exchange Room*);

- `check_access_rights`: this predicate is located of external agents are received. In CARREL 2.0, this is mainly done in states a1 (*Reception Room*), a9 and a10 (*Consultation Room*);

Figures 6.10, 6.11, 6.12, 6.13 and 6.14 depict, in the conversational graphs for each scene, the relations identified. For clarity reasons, not all the relations but only the main ones are included.

Once the relation among the predicates and the states is done, we can check now if the rules (identified in the Rule Level) are met by CARREL 2.0's conversational graphs in its performative structure. However, we will not check:

- rules defining violation situations, as they are handled through the violation mechanisms (violation lists plus agents responsible of violation checking).

- rules about security that are already covered by the Operating System, the database or the agent platform,

- rules that are already covered by the implementation decisions (such as not using information about race and sex).

Therefore, we will only check the precedence rules that are not included in any of the three cases above mentioned:

- **Organ and tissue allocation policies:**

  - `ensure_compatibility` before `assign` (CRL:R3.1, CRL:R4.1): compatibility of pieces is only done in the *Organ Exchange Room* (see figure 6.12) and the *Tissue Exchange Room* (see figure 6.11) scenes, while the final assignation of such pieces is performed later[20] in the *Confirmation Room* scene (see figure 6.13).

  - `plan_delivery` before `deliver` (CRL:R10.1, CRL:R11.1): the precedence relation between these predicates is clearly depicted in figure 6.13.

  - `get_quality` before `assign` (CRL:R1.1, CRL:R2.1): here we refer to a predicate (`get_quality`) which cannot be located in CARREL 2.0 performative structure, as it is a new requirement which emerged from the careful analysis of the regulations made through the HARMON*IA* framework. Therefore, we have to introduce this new requirement. Rules CRL:R1.1 and CRL:R2.1 only state that the quality of the piece must be obtained before the piece is assigned, but they do not specify exactly a moment in time to get the quality information. In ISLANDER this time indetermination is hard to represent in the conversation graphs, as either:

    * We fix a point where the quality of the piece should arrive (e.g., at the *Reception Room*), waiting until such information is available. This solution is not time-efficient, thought.

---

[20]The order between scenes is represented in CARREL 2.0's *performative structure* (see figure 2.8 in Section 2.5.1).

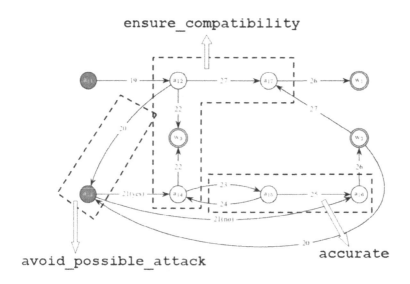

Figure 6.12: Some links from the predicates to the Organ Exchange Room.

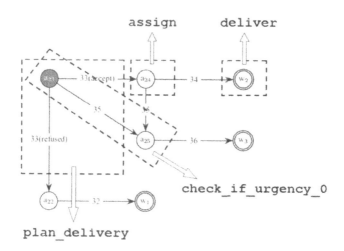

Figure 6.13: Some links from the predicates to the Confirmation Room.

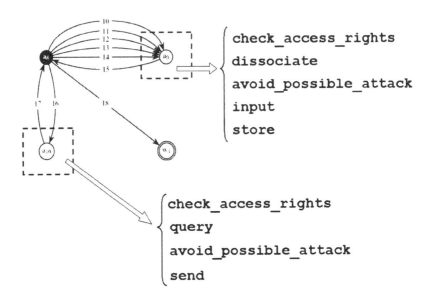

$$
\left\{
\begin{array}{l}
\texttt{check\_access\_rights} \\
\texttt{dissociate} \\
\texttt{avoid\_possible\_attack} \\
\texttt{input} \\
\texttt{store}
\end{array}
\right.
$$

$$
\left\{
\begin{array}{l}
\texttt{check\_access\_rights} \\
\texttt{query} \\
\texttt{avoid\_possible\_attack} \\
\texttt{send}
\end{array}
\right.
$$

Figure 6.14: Some links from the predicates to the Consultation Room.

* We introduce the message in several points of the conversation graphs. This solution is not correct, as it is hard to ensure that we have represented all the points in the interactions where information about the piece quality may arrive.

A solution is to 1) modify the *Reception Room*, which will receive a new agent at anytime with information about the quality of the piece, and then 2) add a rule, to be checked by the *Confirmation Room Manager*, to ensure that positive information about the quality of the piece has arrived before assigning such piece:

* Extension 3: extend the Reception Room conversation graph to allow hospitals and tissue banks to send information about the quality of a given piece anytime, and then check that such information has arrived before doing the final assignation of that piece.

– check_if_urgency-0 before assign (CRL:R16.1, CRL:R17.1): the precedence relation between these predicates is clearly depicted in figure 6.13.

– assign before deliver (CRL:R19.1): the precedence relation between these predicates is depicted also in figure 6.13.

• **Security policy**:

– dissociate before store (CRL:R20.3, CRL:R21.3): In figure 6.14 we can see that, in this case, both predicates are related to a single state, as such predicates describe internal processes (of the agent enacting the *Consultation Room Manager* role) which do not have any related utterance that can be depicted in the conversational graph. Therefore, compliance of rules CRL:R20.3 and CRL:R21.3 is not ensured by the protocol but by the definition of the *Consultation Room Manager* role.

– `avoid_possible_attack` before `send` any data through the network (CRL:R22.5, CRL:R23.5, CRL:R24.5): In this case we again refer to internal processes to be performed by *CARREL* agents, so they are not reflected in the conversation graphs. Therefore, compliance of rules CRL:R22.5, CRL:R23.5, CRL:R24.5 should be included as part of the definition of the following roles:
  * *Consultation Room Manager* role: compliance should be ensured before sending the results of queries which include information about patients (donors or recipients).

  * *Organ Exchange Room Manager* role: compliance should be ensured before sending an organ offer to the hospitals.

– `check_identity` before `access` (CRL:R31.1): the precedence relation between these predicates is clearly depicted in figure 6.10.

– `check_access_rights` before `input` (CRL:R32.1): This precedence relation occurs in two scenes:
  * *Reception Room* scene: as mentioned before, `check_access_rights` is done in state a_1, while `input` is performed later, in states a_3 and a_7 (see figure 6.10).

  * *Consultation Room* scene: in this case, as depicted in figure 6.14, both predicates happen inside state a_9, so precedence is not ensured by the conversation graph but by compliance of rule CRL:R32.1 by the *Consultation Room Manager* role.

– `check_access_rights` before `query` (CRL:R33.1): This precedence relation occurs only inside the *Consultation Room* scene. Again in this case, as depicted in figure 6.14, both predicates happen inside a single state ($a_1 0$), so precedence is not ensured by the conversation graph but through the compliance of rule CRL:R32.1 by the *Consultation Room Manager* role.

– `record` after `access`, `input`, `query` or any incident (CRL:R34.1, CRL:R35.1, CRL:R36.1, CRL:R37.1): these are internal activities to be performed by each scene manager on any relevant event happening inside their scenes. In the CARREL 2.0 prototype this is not implemented yet, so this is a new requirement to be fulfilled:
  * Extension 4: create an access log and an incident log, and modify *CARREL*'s internal agents to perform the recording of any important event in these logs.

  - access before input or query (CRL:R38 CRL:R39): access is done inside the *Reception Room* scene (see figure 6.10) and it is always before any input in the system:
    * inputs inside the *Reception Room* scene: figure 6.10 clearly depicts the precedence relation;

    * inputs or queries inside other scenes: As CARREL 2.0's performative structure imposes that any external agent has to pass through the *Reception Room* scene before entering in any other scene, access control is always done before any other input or query done in the rest of CARREL 2.0 scenes.

An Extension of CARREL 2.0 as *CARREL*'s Procedure Level

In summary, by connecting the specification of CARREL 2.0 with all requirements imposed by the rules in the Rule Level, we have been able to show that the specification and implementation of CARREL 2.0 presented in Section 2.5 is *legal* (in the sense that it follows the regulations imposed by ONT and also the regulations imposed by the context). Therefore, with this we have validated the CARREL 2.0 prototype, as it complies with almost all the requirements that were identified by analysing *CARREL* and its surrounding context through the HARMONIA framework.

The analysis made also identified the extensions needed to make CARREL 2.0 fully compliant with all the rules in the Rule Level and the general implementation decisions in the Procedure Level:

- Extension 1: add inside the CARREL 2.0 platform the data structures representing the rules and the violations, so a) internal agents can reason about them, and b) the external agents receive such rules and violation definitions once they enter into the system.

- Extension 2: add inside the CARREL 2.0 platform the time control mechanisms, by means of timers to detect that a certain time limit has expired.

- Extension 3: extend the Reception Room conversation graph to allow hospitals and tissue banks to send information about the quality of a given piece anytime, and then check that such information has arrived before doing the final assignation of that piece.

- Extension 4: create an access log and an incident log, and modify *CARREL*'s internal agents to perform the record of any important event in these logs.

Therefore, such extension of CARREL 2.0 prototype is a valid implementation of the *CARREL* system.

6.6 Summary

This chapter presents a quite exhaustive example of application of the HARMON*IA* frame-work (presented in Chapter 4 and Chapter 5) to the organ and tissue allocation problem. We have shown how we can derive the requirements of an *e*-organization to support the decision making of a real organization such as the Spanish National Transplant Organization (ONT), by introducing in our analysis not only the requirements imposed by ONT statutes and internal regulations but also the requirements imposed by the Spanish Health System (SpNHS), Spanish Law (SpLaw) and European Regulations (EULaw), which are the context of ONT.

Applying HARMON*IA* we have also shown that CARREL 2.0, the implementation of the *CARREL* system presented in Section 2.5, is *legal* (in the sense that it follows ONT regulations and the regulations imposed by the context), even though we modified the protocols ONT uses in order to optimize the allocation procedure. Therefore, we have been able to address the three major problems we identified in Section 2.6 when using existing approaches and frameworks to highly regulated environments:

P1 *Delegation of responsibility from agents to its owners*: in a framework such as IS-LANDER the only entity responsible of an agent's actions is the agent. However, as we saw in Section 2.5, in CARREL 2.0 we needed to express delegation of all non-fulfilled commitments of the incoming agents to the hospitals they belong to. We informally described the situation with the following rule:

> *IF an agent a_i is representing hospital$_h$,*
> *AND a_i accepts a piece$_n$ in the Confirmation Room,*
> *THEN another agent a_j representing hospital$_h$ must come back to the Consultation Room to update the database about the evolution of the recipient of piece$_n$*

We also mentioned that the delegation of responsibility that is implicit in that rule hardly can be principled by a relation such as *hospital $\succeq$ hospital_finder*, as delegation of responsibility should go from *hospital* to *hospital_finder* and not the other way around. In HARMON*IA* we have solved such problem by assigning, in the Abstract, Concrete and Rule Levels, responsibilities to the super-roles (such as *hospital*), which are then distributed, in the Procedure Level, to sub-roles (such as *hospital transplant coordinator* or *surgeon*) through *part-of* or *power* relations, but always keeping the super-roles into the rules language:

$$CRL:V18.1 \quad done(deliver(organ, hospital_1, hospital_2)) \wedge is\_time(t)$$
$$\wedge wait\_for(is\_time(t + 1\ day)$$
$$\wedge \neg done(hospital_2.send\_result\_transplant(recipient))$$
$$CRL:V18.2 \quad done(deliver(organ, hospital_1, hospital_2)) \wedge is\_time(t)$$
$$\wedge wait\_for(is\_time(t + 3\ months)$$
$$\wedge \neg done(hospital_2.send\_evolution\_3months(recipient))$$

$CRL:V18.3$ $done(deliver(organ, hospital_1, hospital_2)) \land is\_time(t)$
$\land wait\_for(is\_time(t + 1\ year)$
$\land \neg done(hospital_2.send\_evolution\_1year(recipient))$

In the example above, the violations check that the hospital should send the information about the organ at the proper time. The concrete sub-role the *hospital* role empowers to perform this task is irrelevant from the e-organization point of view.

P2 *Design of the protocols*: We also argued that, in highly regulated environments, it is hard to create new protocols or change existing ones while ensuring that they are acceptable. This is caused because, while *norms and regulations* are usually too abstract, *protocols* are too concrete to perform a direct translation from ones to the others. For this reason in the HARMONIA framework we propose a progressive refinement process, from the values to the norms to the rules, which gives as a result the minimum constraints that the regulations and the domain impose to the protocols. Then it is easier to create protocols in terms of efficiency while checking that the imposed rules are observed.

P3 *Check the protocols' compliance of regulations*: An even harder task, which is also related to the gap between the abstractness of norms to the concreteness of protocols. HARMONIA's multi-level approach eases the checking of compliance from one level to the next one. Although we have not presented a full formal connection among the levels, some of our future work will focus in the creation of model checkers that will check, from one level to the next, the consistency of the model.

Chapter 7

Conclusions

In recent years, Multi-agent systems have become an increasingly important approach to implementing distributed applications in complex and real domains such as *e*-commerce, *e*-business, *e*-government or *e*-care. These domains are characterised by the fact that they must support interactions between many different agents that have been developed independently and controlled by different individuals/organisations. The design of open systems in such domains must therefore address a number of difficult challenges, such as the heterogeneity (in architectures, strategies and inner goals) of the incoming agents, incompatible assumptions and limited trust between agents. As we explained in Chapter 1, this scenario has lead to some researchers to introduce social abstractions from human and animal societies in order to solve the aforementioned issues. In most human societies, this social order is achieved by developing norms and conventions that define the expected behaviour of the members of the society. In the case of norms, enforcement is usually supported by some kind of social institutions, which monitor the behaviour of the members of the society, detect situations where behaviour of one or several individuals becomes socially unacceptable and impose some kind of corrective measure.

This book discusses the main issues surrounding the implementation of norms in agent-mediated institutions. In Chapter 2 we introduced the problem of the Organ and Tissue Allocation, and tried to model this highly regulated domain by means of ISLANDER, an existing formalism for *e*-institutions. By doing so, we identified some problems on norm expressiveness that arised during the process. Then in Chapter 3 we analysed the concept of *norm*. By reviewing current work on the field we saw that current approaches are either too *theoretical*, focused on norm formalization by means of very expressive logics that are computationally hard, or too *practical*, focused in the implementation of *e*-institutions but loosing accuracy and expressiveness in the normative system. Another issue is the variety of terminologies (i.e., different terms for the same concept, or the same term with different meaning) that are in use. Finally, an additional issue is that, despite the fact that a significant amount of theoretical and practical work has been carried out in *e-institutions*' definition, formalization and implementation, they have been applied to small or quite simple experimental setups.

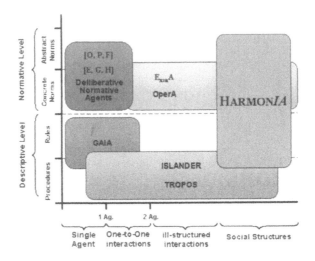

Figure 7.1: HARMON*IA* compared with other approaches.

Our main observation is that norms are specified in regulations that are (on purpose) at a high abstraction level. The level of abstraction is high in order not to be dependent on a circumstantial implementation of the norm. Norms should be stable for many situations and for a relative long time. Therefore, it is obvious that the norms do not have concrete handles for their implementation. In order for the norms to be implemented we presented in Chapters 4 and 5 a formalism in which we can explicitly specify *how* the norms are translated into concrete norms applied in the context of a concrete organization. Using these concrete norms we then translate them in operational representations, such as rules or procedures, to indicate *how* norms are to be implemented in the *e*-organization.

7.1   Our Proposal of a New Framework

Based on these ideas we present HARMON*IA*, a framework to fill the existing gap between formal specifications of norms and their final implementation in rules and procedures (see figure 7.1). It identifies the different levels of abstraction that compose the normative framework, which are the following:

- An *abstract level*, where the values, objectives and context of the *e*-organization are defined (by its statutes) and then translated into *abstract norms*.

- A *concrete level*, where abstract norms are refined into more concrete norms in order to lower the level of abstraction of the normative system by fixing the interpretation of some predicates and terms in the context of the *e*-organization,

- A *rule level*, where concrete norms are translated into rules to be computed by agents. This translation is needed because, as explained in Section 3.2.1, norms have no operational semantics (they only define what *ought to be*, but not *how to be done*).

- A *procedure level*, where the final mechanisms to follow the rules are implemented. We propose to both use rule interpreters to be able to reason about the rules, and to translate the rules into procedures. With this dual approach we aim to cope with two issues:

  - *heterogeneity of incoming agents*: agents entering into the *e*-organization can either follow the protocol, the rules, or both, depending on their capabilities (determined by their inner architecture). Doing so we reduce our assumptions on the internal architecture of the incoming agents, as we only assume that, at least, they will be able to follow the specified protocols.

  - *autonomy and efficiency*: usually, while designing a multi-agent system for a complex domain of significant complexity (that is, a domain that requires a large amount of rules to be properly modelled), the designer(s) should choose between a) deliberative approaches, where agents should reason about all the possible consequences of an action, or b) more reactive, efficient approaches (in time and resources), where interaction between agents is fixed by a narrow protocol that eases the burden of agents when reasoning about the next action to be done. While the former gives autonomy to the agents to decide their actions, the latter greatly reduces the agents' ability to act autonomously since protocols generally give a only very limited choice of actions at any one step. Therefore, there is a conflict between autonomy of the resulting agents *versus* efficiency. In our case we propose to have *Flexible Normative Agents*, which are able to choose between a deliberative or a reactive approach depending on their resources and their time constraints.

In Chapters 5 and 6 we described the process to build an agent-mediated normative electronic organization, from the statutes of the organization to the norms to the final implementation of those norms in rules and procedures. In the case of norms, we use a Deontic Logic that is temporal, relativized and conditional, i.e., an obligation to perform an action or reach a state can be conditional on some state of affairs to hold, it is also meant for a certain type (or role) of agents and should be fulfilled before a certain point in time. For instance, a norm such as *"The National Transplant Organization (ONT) should ensure that the organ is in proper condition before being assigned to a recipient for transplantation"* can be formalized as:

$$O_{ONT}(ensure\_quality(organ) < do(assign(organ, recipient)))$$

In the case of rules, we have chosen a Propositional Dynamic Logic, as it is more suitable to express the operational aspects of organizations, including actions and time constraints:

$$[assign(organ, recipient)] \; done(ensure\_quality(organ))$$

$$\neg done(ensure\_quality(organ)) \rightarrow [assign(organ, recipient)] \; V$$

The first is a *precedence rule*, expressing precedence relations between actions, while the second one is a *violation rule*, expressing a state where the actions of a given entity become illegal and trigger a violation.

The translation steps from one level to the following are described in a formal way, as we aim to be able to verify if a given *e*-organization complies to all the norms that are specified in the regulations.

We also presented those vertical elements that are present in our framework:

- The *policies*, which are vertical solutions related to one or more *values* present in the abstract level. Policy definition comprises different levels of abstraction, from the abstract norms in the abstract level, to their refinement in *concrete norms* and *rules*. The resulting specification of the policy is then translated into the procedures (e.g., triggers, protocols, etc.) that compose the *implementation of the policy*.

- The *role hierarchy*, which is defined by a separation of duties process that starts from the social goals defined in the statutes of the *e*-organization as *objectives*. The role definition not only is used to define the *social structure* of the organization, but also in the refinement of the normative framework, as roles are introduced in the norm definition to reduce their abstractness.

- The *ontologies* of the *e*-organization, which define a) *terms*, *predicates* and *actions* that appear in the abstract, concrete and rule level (*domain ontology*), and b) *performatives* to be used in the communication between agents in the procedure level (*communication ontology*).

- The surrounding *context* of the *e*-organization, which influences the definition of the ontologies and the normative framework at different levels of abstraction.

- The *background knowledge*, a repository where designers can fetch solutions to be applied on the different levels of an *e*-organization specification and implementation.

Our framework is specially suited for those complex, highly regulated domains where the behaviour of a real organization or an *e*-organization has to follow regulations that define restrictions at different levels of abstraction. In order to explore this problem, we moved out of the widely studied *e*-commerce field – which often can be modelled by the definition of interaction rules based in game-theory) – to explore scenarios with more complex and less well understood normative frameworks. The case study we chose (the human organ and tissue allocation problem) is a good example of a scenario highly regulated by national and international laws. We have been able to refine our framework in order to cope with similar problems in the context of electronic Health Care (*e*-care) and *e*-government, which will be the application areas that we will address in our future work.[1]

7.2 Original Contributions

Our work presents both theoretical and applied contributions to the field. In the case of the theoretical contributions, we can summarize them in: a) distinction between normative and operational, b) the relation between normative systems and contexts, and c) introduction of new terminology. In the case of applied contributions, we can summarize them in two: a) the connection between formal specification and agent implementation, and b) norm enforcement as detecting illegal worlds.

7.2.1 Distinction between Normative and Operational

One of the novel contributions is that we have remarked the distinction between the normative level of the e-organization (the one that states what is acceptable or unacceptable by means of *obligations*, *permissions* and *rights*) and the practical, descriptive level defining the *praxis*, that is, how agents should behave in order to meet the norms. In the former, values and norms stating are placed, while in the latter, rules and procedures describe how the norms are to be interpreted and used by the agents. Doing so we propose an alternative to previous approaches, which tried to use the same representation or formalism to model both aspects, either by extending deontic-like formalism with action logics, or by extending operational formalisms with some deontic concepts. In this work we propose the use of three different representations (Propositional Deontic Logic, Dynamic Logic and any procedural language), each one fitted to describe a normative framework in a different level of abstraction.

7.2.2 Normative Systems and Contexts

We have studied the relation of normative framework and contexts. We have identified a normative system as a context (definition 4.7) which defines a set of (consistent) norms and a shared vocabulary that apply within the boundaries of the context. We have also presented the idea of nested contexts in order to model the influence of the environment of a given organization. We have seen how a sub-context fixes the interpretation of the norms stated in the super-contexts, reducing their abstractness, and we propose the use of a *counts-as* operator ($\Rightarrow_s$) to formally model such interpretations. We have also studied how the super-contexts can affect the definition of the normative framework in different levels of abstraction (from values and norms to the final rules and procedures). In order to represent the complex relation between normative systems, contexts and the agents'

---

[1] The European Union FP6 program explicitly mentions *e*-care and *e*-government as being part of the priority areas of future research, as stated by the FP6 1.1.2.i priority defined in their list of priority thematic areas.

beliefs and goals, we also present here a graphical notation in order to easily represent the different levels of the normative system and their effect in the agent's beliefs and goals. [2]

7.2.3 New Terminology

In this book we provide new definitions for concepts such as *institution, e-institution, organization, e-organization, norm* (abstract and concrete), *rule, procedure, role, policy* and *context*. We want to remark the distinction we make between *institutions* (as abstract, quite non-contextual patterns) and *organizations* (as concrete, contextual instances of one or several institutional patterns).

7.2.4 Connection between Formal Specification and Agent Implementation

The HARMON*IA* framework proposes a formal connection between the different representations used to model the normative framework:

- *from Deontic Logic to Dynamic Logic*: by using Meyer's translation rule [138]:

$$F(\alpha) \mapsto [\alpha] V$$

- *from Dynamic Logic to Procedural Languages*: by the existing connection among both representations expressed in the following axiom:

$$\{P\}\alpha\{Q\} \equiv P \rightarrow [\alpha]Q$$

This connection is very useful not only in its *top-down* direction, but also from *bottom up*:

- *top-down*: the connection from norms to rules to the final procedures guides the design process of the *e*-organization, as it identifies the minimum set of restrictions that are defined by the normative framework, and eases the task of checking if the implemented procedures follow such restrictions.

- *bottom-up*: agents can trace the origin of a given protocol and reason in terms of the rules and norms the protocol implements. This allows the definition of *Flexible Normative Agents*, which are able to handle those unexpected situations that a given protocol has not considered, reason in terms of the related rules and adapt their behaviour appropriately.

---

[2] An example of this graphical notation is figure 7.5, which allows us to represent relations between contexts.

7.2.5 Norm Enforcement as Detecting Illegal Worlds

In our framework we propose that enforcement of norms should not be made in terms of direct control of a central authority over the goals or actions that the agents may take, but through the detection of the *violation states* that agents may enter into and the definition of the *sanctions* that are related to the violations. With this approach we do not make strong assumptions about the agents' internal architecture, as the *e*-organization only monitors the agent behaviour (that is, agents are seen as *black boxes*.). The enforcement of the norms in an *e*-organization is achieved through a special kind of agents, the *Police Agents*, which monitor the behaviour of the agents, detect violations and check the compliance of the sanctions.

7.3 Ongoing Work

There is some ongoing work that, because of its preliminary state, has not been included in previous chapters of the book. However, because of its relevance, in this section we present some outlines on our current research in the following areas:

- Definition of a (generic) modular architecture for *e*-organizations.

- Creation of tools for *e*-organizations.

- Testing the framework in new domains

7.3.1 Definition of a Modular Architecture for E-Organizations

The HARMON*IA* framework presented in this book is a framework centered on the normative aspects of institutional modelling, and it is suited for highly regulated domains where the design and the final behaviour of the multi-agent system should meet a corpora of norms and rules. However, *e*-organizations can be also applied to less normativized domains where design should be guided by organizational issues. Therefore there is a need of:

- A modelling framework for different types of MAS, from closed systems to open, flexible environments, and from environments with no norms to highly regulated domains.

- A modular architecture, identify all facilitators and structures needed inside a MAS, depending on the characteristics of the domain. The final objective is to create templates[3] that can be adapted, parameterized or instantiated to build an *e*-organization.

With these needs in mind, we are defining a new framework called OMNI, integrating the normative aspects of HARMON*IA* with the organizational aspects of the OperA framework[58]. OMNI aims to be a framework for modelling a whole range of MAS, from closed systems with fixed participants and interaction protocols, to open, flexible

---

[3]In our terminology these templates are called *e*-institutions (see definition 5.10).

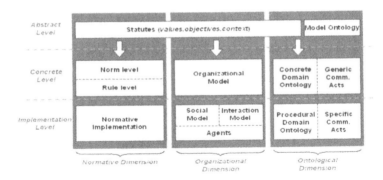

Figure 7.2: The OMNI framework.

systems that allow and adapt to the participation of heterogeneous agents with different agendas. This approach is rather unique, as most existing frameworks concentrate in a specific type of MAS.

OMNI is composed by three dimensions: *Normative, Organizational* and *Ontological* that describe different characterizations of the environment. The Normative Dimension is an extension of the normative levels in HARMON*IA* , while the Organizational and Ontological Dimensions are extensions of concepts and models in OperA. Figure 7.2 depicts the different modules that compose the proposed framework, organized into three levels of abstraction:

- the *Abstract Level*: where the statutes of the organization to be modelled are defined in a high level of abstraction. This step is similar to a first step in the requirement analysis. It also contains the definition of terms that are generic for any organization (that is, that are incontextual) and the ontology of the model itself.

- the *Concrete Level*: where all the analysis and design process is carried on, starting from the abstract values defined in the previous level, refining their meaning in terms of norms and rules, roles, landmarks and concrete ontological concepts.

- the *Implementation Level*: where the design in the Normative and Organizational dimensions is implemented in a given multi-agent architecture, including the mechanisms for role enactment and for norm enforcement.

Different domains have different requirements concerning normative, organizational and communicative characteristics, which means that not always all three modules have the same impact or are even needed: in those domains with none or small normative components, design is mainly guided by the Organizational Dimension, while in highly regulated domains the Normative Dimension is the most prominent.

As in HARMON*IA* , the modelling process in OMNI starts with the *Statutes*, which indicate in a very abstract level a) the main *objectives* of the organization, b) the *values* that direct the fulfilling of this objective and c) the *context* where the organization will

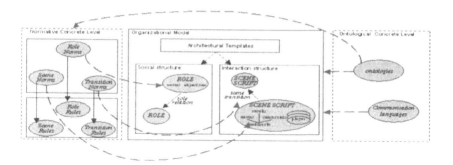

Figure 7.3: The Organizational Model and its interactions with the Normative and Ontological Dimensions.

have to perform its activities. The *objectives* of the organization express the overall goals
of the society, and are the root of the Organizational Dimension. *Values* define beliefs
about, for instance, what is acceptable or unacceptable, good or bad, and they are the
basis of the *Normative Dimension*.

The Organizational Dimension

In OMNI, the Organizational Dimension consists of a 3-layered model: based on the
concerns identified in the Abstract Level, the Concrete Level specifies the structure and
objectives of a system as envisioned by the organization, and the Implementation Level
describes the activity of the system as realized by the individual agents. This separation enables OMNI models to respect the autonomy of individual agents while ensuring
conformance to organizational aims. The organizational dimension is composed by three
interrelated models that come from OperA (already depicted in figure 3.1):

- The *Organizational Model* specifies the organizational characteristics of an agent
  society in terms of social structures (roles) and interaction structures (scene scripts).
  This model specifies the organizational characteristics of an agent society by means
  of the following components:

    - The *Social structure*, which specifies, from the objectives, the description of
      relationships and capabilities of roles, role groups and activities.

    - The *Interaction structure*, indicates the abstract processes that according to the
      organization's view must be used to achieve its objectives. This description is
      declarative in nature, describes the states (*landmarks*) that agents must strive
      to achieve, instead of the activities to perform.

    - The *Architectural Templates*, which describe three kinds of patterns (*hierar-
      chy*, *network* and *market*) that, if applicable to a domain, support the definition
      of the organizational structures (see table 7.1).

	Market	**Network**	**Hierarchy**
Agent 'values'	Self interest	Mutual interest/ Collaboration	Dependency
Coordination	Price mechanism	Collaboration	Supervision
Dependency relation	Bidding	Request	Delegation
Facilitation roles	Matchmaker, Reputation facilitator, Market master	Matchmaker, Gatekeeper, Notary, Monitor	Controller, Interface facilitator

Table 7.1: The three architectural templates.

- In the *Social Model*, the enactment of roles by agents is fixed in social contracts that describe the capabilities and responsibilities of the agent within the society, that is the agreed way the agent will fulfil its role(s).

- In the *Interaction Model* concrete interaction scenes are dynamically created by role-enacting agents, based on the interaction scripts specified in the OM. Role enacting agents negotiate specific interaction agreements with each other and fix them in interaction contracts.

Contracts are introduced as a means to integrate top-down specification of organizational structures with the autonomy of the participating agents. These contracts must specify the activity of agents as enactors of society roles, and include aspects such as the specification of the role(s), the time period the contract holds, specific agreements and conditions governing the role enactment, and the sanctions to take when norms are violated (especially if specific sanctions are agreed upon).

The Normative Dimension

The normative dimension is an adaptation of the HARMON*IA* framework to the models in the Organizational Dimension. The main changes from the framework presented in this book are the following:

- Norms and rules are included in the definition of the roles, the role groups and the scenes.

- A new relation between roles, the *dependency relation*, has to be taken into account in order to express normative influence.

- The connection between the rules' predicates and the final states of the interaction protocols is done through the *landmarks* that describe such interaction in a higher level of abstraction. Therefore, the landmarks are a middle layer between the rules and the final procedures.

- Agents may negotiate some clauses for the enactment of a role by means of the contracts. These clauses may introduce (negotiated) changes in the norms and rules that apply to the agent.

The Ontological Dimension

In OMNI, the Ontological Dimension describes both the content and the language for communication, at three different levels of abstraction. At the Abstract Level, the *Model Ontology* can be seen as a meta-ontology that defines all the concepts of the framework itself, such as norms, rules, roles, groups, violations, sanctions and landmarks.

The content aspects of communication, or domain knowledge, are specified by *Domain Ontologies*. In OMNI abstract concepts can be iteratively defined and refined in terms of more concrete concepts. The Concrete Domain Ontology includes all the predicates and elements that appear during the design of the Organizational and Normative Structure, and the Procedural Domain Ontology, with the terms from the domain that will be finally used in the implemented system. Concepts or predicates at a lower abstraction level, count as, or implement, concepts at the higher levels. For instance, the actual realization of the AAMAS'04 conference *counts-as* the IFMAS's objective *organize-conference* defined in the Organizational Model, which in turn *counts-as* the IFMAS's value of disseminate knowledge, described in its statutes.

Communication Acts define the language for communication, including the performatives and the protocols. At the Concrete Level, *Generic Communication Acts* define the interactions languages used in the Organizational Model, while the *Specific Communication Acts* covers the communication languages actually used by the agents as they agree in the interaction contracts. As with the content ontologies, communicative acts defined at a lower level of abstraction implement those defined at a higher level.

A first report on the different components of the OMNI framework can be found in [215].

7.3.2 Creation of Tools for E-Organizations

In order to ease the design, implementation and verification of *e*-organizations, new tools are needed in order to:

- model those aspects of the surrounding context that may impact in the *e*-organization;

- assist in the refinement process, from the abstract values to the rules, by ensuring that the links between levels are consistent, and suggest, at each level of abstraction, the influence from the context (values, norms, rules, even complete procedures to follow);

- link the specification with existing domain ontologies defining terms, predicates and actions;

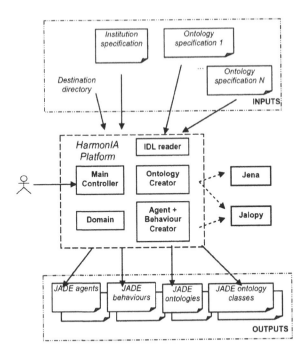

Figure 7.4: The Architecture of the current HARMON*IA* platform (source: [164]).

- create the background knowledge repository, in order to be able to re-use previous solutions during the specification of new *e*-organizations, including templates for policies;

- include model-checking functionalities, in order to check a) the consistency of the specification at each level, and b) the compliance by a given level of the restrictions defined in higher levels.

- tools for (automatic) generation of protocols: the use of planners to create protocols that follow a set of rules in the rule level.

With all these objectives in mind, we have started the development process of the HARMON*IA* suite of tools. In its first phase the HARMON*IA* platform is composed by two main components:

- The *Institution Management Framework*, which runs on top of a JADE platform [16] and is composed of a set of internal agents performing control and validation of the *e*-organization at run-time.

- The *Institution Generator*, which generates agent source code and supporting files from an institution specification.

The current implementation and the language used to define *e*-organizations are very close to ISLANDER, and only cover part of the rule level and the procedure level. The main difference between our approach and ISLANDER is that there are no *governors* between the *external agents* and the *scene managers*. During the next phase we will introduce concepts such as *landmarks*, *violations* and *sanctions* in order to make the tool fully compliant with the rule level in the HARMON*IA* framework. Further phases will add other tools to create the normative specifications and to perform model checking on the different levels of an *e*-organization model.

More details on the HARMON*IA* tools can be found in [165][164].

7.3.3 Testing the Framework in New Domains

We are currently applying the HARMON*IA* framework to an *e*-government scenario. The police organizations in Netherlands (as it happens in other countries) manage and use an increasing number of distributed electronic databases. The use of these databases is governed by Dutch and European privacy regulations. In many cases these regulations are supplemented by local rules defined by local councils, rules to be followed by the departments that manage the individual databases. As a consequence, a complex system of contextual, multi-level rules has emerged that is characterized by a mixture of (international and national) norms and a great diversity of (formal and informal) local rules. In the present situation this legal knowledge is mainly applied depending on the knowledge of police officers that introduce, access and maintain the information in the databases.

The ANITA project (Administrative Normative Information Transaction Agents) involves several Dutch research groups[4] and aims to a) investigate on proper methods for the acquisition of norms, b) design and build an agent framework that enhances the application of norms, and c) test a prototype in practice with members of the police and judicial intelligence. In this framework, facilitator agents will decide, based on norms, whether to allow the access or the share of concrete pieces of information.

More information can be found at the project website [5].

7.4 Suggestions for Further Research

As a result of this work, we envision some promising future lines of research:

- **Extension of the normative framework**: extend the framework, mainly in two directions:

  - In our framework we only consider sanctions that are related to a single violation. But the framework should be extended to include those cases where there are special sanctions to be applied by the occurrence of several violations.

---

[4]The ANITA project is a multidisciplinary joint venture of legal and artificial intelligence departments of four universities (University of Groningen, University of Leiden, University of Maastricht, University of Utrecht).

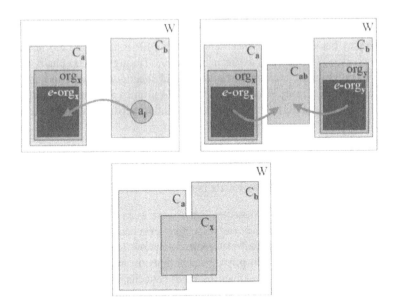

Figure 7.5: Three issues on Normative Dynamics. 1- (top-left) *Change of Context*: an agent passing from one context to another. 2- (top-right) *Consensus*: two e-organizations defining a context with shared interpretations. 3- (bottom) *Collision*: defining a sub-context which inherits from more than one context.

  – Formalize normative influence dynamics by defining axioms to describe how normative influence spreads through the *power*, *is a* and *part-of* relations defined between roles, and study the relation of normative influence with related concepts such as *delegation, authorization, responsibility* and *representation*.

• **Study the use of other languages**: In this work we proposed the use of three different representations for different levels of the normative framework (Deontic Logic, Dynamic Logic and Procedural languages). We chose these representations because there were already in literature formal connections defined between them that we used to connect formal specification with agent implementations. But further research can be done in order to find other combinations of languages to be used at each level in order to be expressive enough and suitable for verification methods.

• **Study the case of precedence-based legal systems**: In Section 3.1.2 we explicitly focused our research on legal systems based in Roman Law (i.e., systems defined by corpora of written norms and rules), as there is quite a lot of work on formal models for rule-based systems. That makes our framework not suitable for those countries with legal systems based in precedence (legality of a current case is defined by past decisions of judges in similar cases). This is the case, for instance, of the Common Law, applied in the United States, the United Kingdom and associated states of the CommonWealth. Current research on the AI field applied to precedence-based systems is focused on applied approaches to the design and construction of systems that are able to compare cases and reason with them. However a formal link between this national (precedence-based) system and the international (rule based) laws should be found. The impact of this change in the HARMON*IA* multi-level framework would be located in the Concrete Level, where part of the norms would now be expressed as cases in, e.g., a Case-Library (that is, in this case we are using a different language $CNorms'$ to express the concrete norms, a language based on instances or cases). Then, in order to reduce the amount of resources that the agents may need in their reasoning cycle, a classification mechanism could generalize the cases by expressing them in rules of e.g., a decision tree. In order to allow verification, a formal definition of $CNorms'$ and its connection with the rules should be properly defined.

- **Study of dynamic aspects of normative systems**: In our framework we have not considered the problems that may arise by conflicting assumptions defined by different contexts and how contexts may play and important role to reach *consensus* on a) norm definition, and b) norm, rule and protocol negotiation. There are three main scenarios that should be studied (see figure 7.5):

  - *Change of context*: how external Deliberative Normative Agents, entering in a context, can cope with the collisions between the agent's values, beliefs and goals (defined in terms of a given context) and the ones defined by the e-organization context.

  - *Consensus among e-organizations*: the achievement of consensus between organizations that are defined in different contexts can be eased by:
    * identifying a super-context shared by both *e*-organizations: by using the shared values and norms as basis for consensus. An example of this is the use of European Law to reach consensus between European organizations.

    * create an intermediate context with the agreed definitions of values, norms and rules to regulate the interaction among both organizations. An example of this scheme is the creation of bi-lateral treatises between countries.

  - *Collision in context definition*: the study of how contexts defined from several, conflicting super-contexts can cope with the conflicting values, norms and terms. The solution will consist in the definition of some kind of multiple inheritance mechanism to solve those collisions (in a similar way as it is done in Object-Oriented analysis and design).

Part III

Appendix and Bibliography

Appendix A

Medical Data Protection and the Internet

> *"Whatsoever things I see or hear concerning the life of men, in my attendance on the sick or even apart therefrom, which ought not to be noised abroad, I will keep silence thereon, counting such things to be as sacred as secrets."*

Hippocrat

This appendix is the result of the analysis made about the laws and regulations that rule in the creation of software medical applications which manage medical information about patients in Spain. This analysis was needed in order to get acknowledged with the requirements by law of a system such as *CARREL*/UCTx.

The main aims of the work were: 1) to analyse the evolution of legislation, both Spanish and European, on personal data protection -and, in particular, clinical data- and 2) to propose mechanisms which will allow the accomplishment of their requirements, and, at the same time, will facilitate the task of an organ/tissue collection unit.

We will also discuss the opportunity of using electronic formats to store case-histories, as a means to speed up the transmission of information and to improve its management, focusing on a general improvement of health services.

A.1 Medicine, Information Technology and Privacy

Protection of individual's health-related data has been a continued concern of the medical body from the very beginning of the medical practice, as reflected in the famous Hippocratic oath. It obliges the physician to conserve the secret of as many information he or she will obtain from the patient during the treatment.

Changes in medical practice, its institutionalization, the appearance of case-histories as a method of collecting important data about the patient which facilitates the task of the physician, etc., have led physicians, health institutions and, in the last century, the health-related industries (insurance companies, pharmaceutical industries, etc.) to the development of different methods to achieve the privacy of the data obtained and stored from patients.[1]

The appearance of computers, first as data storage systems -giving rise to large medical information archives– and, secondly, as information sharing and management systems, has led to a true revolution in the design of the methods needed to secure medical data information. And, in the middle of all that agitation, a second wave in this revolution happened with the advent of the Internet, as a means for communicating and sharing medical information, including its impact in medical practice (e.g., *telemedicine*).

The use of information technology for the encoding of medical data, the possibility of relating medical data from several sources (for example, the databases of several European health centers, or the Icelandic databases which would allow even to trace the genetic history of that island's population) and the fact that this information might be exploited by a company outside the scope of health (e.g., a life insurance company) have created distrust among users, which are in fact the legal owners or their personal data and who want to control the diffusion of these data, especially medical data.

It is possible to point out two types of concerns derived from the use of computers to process medical data: a) possible *leaks or filtering* of information caused by individuals or institutions, and b) possible *diversion* of information from health institutions to third-parties in order to exploit it. This has led to a plethora of policies, practices and procedures for the processing of *ad hoc information*, which have derived in *de facto* standards quite difficult to arrange. However, we must recognize that, before the advent of computers, security and privacy of medical data already were quite remarkable problems in developed societies, where health services reach the majority of their populations.

The states have tried to regulate the protection of personal data to avoid, as far as possible, its improper use and diffusion. In Medicine, it is evident the accelerated use of information and communication technologies to communicate and share medical data. Today, the use of computers is generalized among the medical body, but this situation must be considered as recent, and is still far from the ideal in the sense of the optimal use of available resources. In parallel to this increase of use, it also exists a rise in the computing hardware and software investments, not only to speed up the exchange, but also to ensure the integrity, confidentiality and security of the information. And, although at this moment the transmission and sharing of medical information - at least in the case of Spain - is limited to the borders imposed by the local network of an institution, it is easy

---

[1]The terms *privacy, confidentiality* and *security* are used in many ways when discussing the protection of personal medical information. The convention used here will be to use *privacy* referring to the desire of an individual of limiting the distribution of his/her existing medical information. The term *confidentiality* will refer to the conditions under which personal medical information is shared and/or distributed in a controlled fashion. *Security* refers to the measures implemented by the organizations in order to protect the information in charge of them and the systems on which it is stored. It also includes the efforts to respect privacy rights, to maintain the confidentiality and ensure the integrity and availability of the information.

to forecast a rapid change of this situation, when institutions and health centers recognize the advantages of sharing information between them.

There exist considerable efforts to put in practice a body of policies which ensure the protection of medical data in a scenario of massive use of computers in the health sector. We can see an example in the United States of America, with the creation of the *Health Informatics Standards Board* (HISB), which is dependent from the *American National Standards Institute* (ANSI). This organization must generate the procedures to create: 1) medical data storage structures in electronic format; 2) structures for the interchange of data, imaging, sounds, and medical signals between medical institutions; 3) coding and terminology suitable for distribution of medical-content messages, 4) communication mechanisms with diagnostic instrumentation and other equipment which produces medical information; 5) design of communication and representation mechanisms for the medical protocols in data and knowledge databases, and 6) mechanisms which ensure privacy, confidentiality and security of medical information.

A.2 Advantages of Electronic Formats for Medical Data

The main benefit of deploying and using electronic formats to collect patient's case-histories lies in the faster access to those data and in an increased and automated control over this access. That is, it is possible to identify genuinely the authorized users and to know the scope of their privileges when accessing the information, as well as the maintenance of an access log as required by law. Moreover, case-histories in electronic format allow the authorized personnel: 1) accessing to the information from different locations, 2) sharing information with other authorized users, and 3) the concurrent access (that is, at the same time) of several people to the same information.

These electronic records allow the maintenance of an individual's case-history throughout his/her lifetime, including current prescriptions, laboratory results, and even many kinds of imaging (radiography, tomography, etc.).

A.3 Requirements to be Fulfilled

The requirements can be divided in two groups:

- requirements coming from the medical community demands: they are mainly concerned to the functionalities an electronic support for medical data might have to be useful for their daily work.

- requirements from current legislation on personal data protection: they are mainly concerned to the security measures to be taken into account when managing personal data.

A.3.1 Requirements from the Medical Community

Multimedia support of medical data should focus on facilitating the task of the physician when a diagnostic is needed. This requires that the specification of those formats agree to a set of very demanding requirements, among which are:

1. *Expressiveness*: a physician should be allowed to shape any information he/she considers relevant, and this information must integrate in a natural way with the rest of them;

2. *Flexibility*: it must allow any member of health personnel to manipulate it without the need of deep computing skills;

3. *Security*: it must give enough reliance to physicians, patients, and to the society as a whole;

4. *Robustness*: it must withstand possible attacks or misuses without getting out of service;

5. *Verifiable*: it must allow an auditing at any time, in order to validate the consistency of the information contained in the system;

6. *Reusable*: the information about a patient generated in a given health center or service must integrate seamlessly in another.

A.3.2 Law-Enforced Requirements

Medical data, such as a patient's case-history, are considered by the current legislation as one of the personal data types requiring a highest protection level. Concerns about the influence of computers on person's intimacy have led many countries to the creation of suitable regulations. In the case of Spain, legislative control of computing activities was introduced from the very beginning in the Spanish Constitution [44], which, in its Article 18.4, calls on the legislator to limit the use of computers to guarantee honor, personal and familiar intimacy of citizens and legitimate exercise of their rights.

But Spanish Constitution only takes into account intimacy, not privacy. In Spanish legislation, privacy stands for a wider set of aspects of an individual's personality, which might be irrelevant or have no intrinsic mean when considered separately, but can generate profiles or portraits of individual's personality when joined together. Organic Law 5/1992 (also known as LORTAD [117]) has been developed, having in mind the precise goal of protecting privacy. It states the aforementioned definition of privacy, and also establishes a whole regulation body on 1) data archives[2] which contain information about individuals,[3] 2) its allowed processing and use, be it automated or not, by public or private entities, 3) obligations related to data quality, technical security and notification to a controlling

---

[2]*Data archive* means any kind of physical media to store information, be it a computer system (computer files, databases), paper, microfilms, etc.

[3]A person which is fully identified by those data, or which can be identified starting from those data. We will refer to that person as the *interested* or the *affected* person.

authority,[4] and 4) the rights of individuals to access those data, request its amendment or even be opposed to its processing.

Personal data are defined as those which allow the identification of a person, and which reveal racial or ethnic origins, political opinions, religious or philosophic beliefs, trade union's affiliation, as well as data related to health or sexuality. Therefore, medical data are protected by this Organic Law.

The LORTAD, including its extension by virtue of Royal Decree 1333/1994 [179], establishes the following general regulations:

1. *Data quality*: data must be exact and up to date. This regulation is not a problem in the case of medical data, which, by its own nature, must be correct at any time.

2. *The person must be informed about his/her rights over the collected information*: the affected persons whose data are requested shall be informed about: a) their data will be stored in an archive, b) the way these data will be processed, c) the right to access, amend, cancel o be opposed to the collection of those data, and d) the identity and address of the responsible of processing the data, who will allow to be contacted by the affected. In the case of medical data, this regulation is modified by Article 8 in the same LORTAD, which automatically allows the institutions and health centers to collect and process information from the patients who come along or are being treated, in accordance with laws such as the *General Health Act* (LGS [114]), *Drug Act* [115] or Organic Law 3/1986 [116]. As an example of modification produced by these previous laws, the LGS states that there is no need of patient's consent in order to obtain data for the treatment, in the event that the no-intervention would be dangerous to public health, or when urgency does not allow treatment delays, which would give rise to irreversible injuries or death risk.

3. *Individual's consent*: as a general rule, personal data processing requires the consent of the affected person. In the case of medical data, in the same way as the previous regulation, this norm is modified following specific laws such as the LGS.

4. *Data communication*: the data can be given up or communicated to third parties only if these parties need the data to accomplish the goals for which the data have been collected, and if the affected person or persons have consented this transference. In the case of medical data, Article 8 above mentioned gives the permission of data processing to any health institution or center, being it public or private; therefore, data transference can be done to any health institution or center in the cases provided in Article 11: *consent* of the affected person (the patient) is not needed when a health-related data transference has to be done, if such a transfer is necessary to resolve an urgency for which access to an automated archive is required, or to develop epidemiologic studies, as stated in Article 8 in LGS. In a similar way, Article 33 in LORTAD permits the international transference of data among practitioners or health institutions.

---

[4]The Data Protection Agency (APD), created by the LORTAD.

5. *Data security*: the necessary organizational and technical measures must be adopted in order to guarantee the security of personal data, thus avoiding its unauthorized access or processing, improper alteration or unrecoverable loss. The applicable security measures were specified by means of Royal Decree 994/1999 [182], which we will deal with in Section A.5.

6. *Notification to Data Protection Agency*: the creation of any archive or physical media to store personal data must be notified to the Data Protection Agency. This notification must specify, among other items: who is responsible of the archive, the intended usage of the data, the structure which will be used to store them, the processing which will be done over them, and the foreseeable transferences of the data. On the basis of this notification, the Data Protection Agency will authorize the creation of the data archive. Any modification in the structure of the data must also be notified, as well as the first transfer of the data.

Until 1994, Spain was not the only state with a legislation on this subject. The European Union (as opposed to the situation at the United States) manifested a special concern about the protection of their citizen's personal data. In 1995 there were already several countries with suitable legislations about processing personal data. However, these legislations did not allow the exchange of personal information between states. European Parliament created the 95/46/CE Directive [62] with the purpose of homogenizing legal cover on data protection, in order to warrant an appropriate protection level on each transfer inside the European Union. That directive [5] did not cause major changes in Spanish legislation, which already had a strong protective law, yet it settled some nuances and regulations that were subsequently added to the new text of the law about data protection: Organic Law 15/1999 [118].

Following are listed the modifications introduced by the European directive:

1. *Extension of the field of application*: The field of application of the legislation is extended to every data archives structured in a way which allows the easy extraction of personal information, irrespective of any kind of automated processing of the information.

2. *Data maintenance*: Personal identity data will only be kept during the period needed to reach the aims they were collected for, or the authorized extensions of those aims. If it is desirable to maintain this information long after this period (for historical, statistical or scientific purposes), it must be done in a way that avoids personal identifications. In the case of medical data, allowed limits are provided in Recommendation R(97)5 [183] (see Section A.4).

3. *Exception to prohibition of processing medical data*: Article 8.3 allows personal data processing *"when it is needed for medical prevention or diagnosis, health care*

---

[5] At the end of year 2000, the European Parliament extended the personal data regulations initiated by this norm by means of Regulation (CE) 45/2001 [185], which covers all that was already established by the Directive 95/46/CE, determines the penalty mechanism at the European level, and creates the figure of the Data Protection European Supervisor as an independent control authority.

or medical treatments, or management of health services, provided that such data processing will be done by a health practitioner under professional secret, be it by virtue of national legislation or the norms established by the competent national authorities, or by another person likewise subject to an equivalent obligation of secret".

4. *Data transference*: The EU member states will not restrict nor prohibit the free circulation of personal data between member states by reasons related to the guaranteed protection level, since the Directive is homogenizing the protection level throughout the whole European Union. In the case of Spain, this norm had little effect, since in the Spanish Order of February 2, 1995 [156], which defined for the first time the list of states with regulations on data protection comparable to the Spanish one, Italy and Greece were the only states of the EU to which data transference was not allowed, because of their lack of regulations on data protection until 1996 and 1997. The Spanish Order of July 31, 1998 [157] already included them in the list.

5. *Data security*: unlike the LORTAD, where the obligation of taking care of data security lies exclusively in the person responsible of the data archive, the Directive states that any other person responsible of the processing of such data will also have to take care of data security, by the application of all the necessary measures.

All the aforementioned modifications were later incorporated in Spanish Organic Law 15/1999 (also known as LOPD). This law completely substitutes the LORTAD, in order to adapt it to the new assumptions of European directives. Apart from the mentioned modifications, the LOPD adds the following changes:

1. *Data source*: public and private access data sources are distinguished separately. *Public access data sources* are those which can be accessed by any person, with or without a fee, but without legal restrictions of any kind. Some examples of this type of sources are telephone directories, newspapers and mass media, or the Spanish State Official Bulletin (BOE). In the case of sources such as books or other physical supports, the obtained data will lose their public character after a new edition appears. In the case of data obtained through telematic networks (i.e., the Internet), these data will lose their public character after one year from the date they were obtained.

2. *Exception in the notification to the person*: when the data come from a public access data source, it is not necessary to notify the person about the intended processing of that information.

3. *Right to be opposed to the treatment*: in the cases in which there is no need of person's consent to initiate a treatment, that person can be opposed to the treatment if no other regulation states the contrary.

4. *Duty of Notification*: Failing to send the law-enforced notifications to the Data Protection Agency (for example, the notification of the creation of any personal data archive) is considered a serious infringement of law.

5. *National data transference*: data transference to third parties will be only allowed if a previous dissociation process is made, thus avoiding the identification of given individuals.

6. *International data transference*: the Data Protection Agency will be informed before the first data transfer to a given country is made. The Agency will determine if the target country has the suitable protection level.

A.4　Privacy and Security of Electronic Medical Data

It is widely accepted, at least in Western World, that there exists a need of protecting the medical information about persons. In the United States, due to the lack of regulations which protect all kinds of personal data, specific regulations on clinical information are being created. These regulations begin with the recommendations from the 1996's Health Insurance Portability and Accountability Act. After a long process, a Privacy Regulation has been defined [96], which has been theoretically put into effect in February 2001, although in most cases a term of 2 to 3 years is granted in order to allow the adaptation to the regulation.

In the case of European states, medical data are already covered by the Directive 95/46/CE, so it might seem unnecessary to define additional regulations. Even so, in 1997 the Recommendation R(97)5 [183] about medical data was drafted, completely substituting a previous recommendation (1981). This recommendation basically arose for two reasons: 1) it was noticed that the advances on medical science strongly depend on the availability of medical data about individuals, and 2) an increase of the use of automated information systems for processing medical data was detected, not only for medical assistance and research, hospital management and public health, but also outside the health sector, which was a reason of concerns.

The R(97)5 makes almost no modifications to the regulations commented above, since it basically translates the regulations of the Directive 95/46/CE to the clinical data language, and, in several norms, it specifies the limits of what can or cannot be done with medical data. Some of the precisions it introduces are as follows:

1. *Collection and processing*: the R(97)5 permits the collection and processing of medical data in the following cases: a) by reasons of public health; b) for preventive medical purposes, or for diagnostic or therapeutic purposes related to the affected person or a relative in his/her same genetic line (in this case, the regulation allows the processing of these data for the purpose of offering a positive medical service to the patient); c) to establish, exercise or defend a legal complaint; and d) to repress a specific crime, or for any other public interest of importance.

2. *Rights of access and amendment of the affected person*: the access of the affected person to his/her medical data may be revoked, limited or rejected, only if it is so stated in the laws, and under the following assumptions: a) by reasons of State or public security, or crime repression, b) the knowledge of some data may be seriously harmful to the health of the affected person; c) the information about the affected

person also reveals information about other persons (for example, consanguineous or genetically close relatives); d) the data are used with scientific or statistical purposes and it is clear that there is no risk of intimacy violation.

3. *Right of cancellation*: the affected person must be informed of the possibility, if any, of denying the consent to the collection and processing of his/her medical data, or taking away an already given consent, and of the consequences of this cancellation.

4. *Preservation*: medical data can only be stored longer than needed for the purpose they were collected for: a) in the legitimate interest of public health or medical science; b) to allow the person who takes care of the medical treatment or the administrator of the data archive to defend himself when faced to a legal complaint; c) by historical or statistical reasons (in this case the suitable measures must be adopted in order to maintain the patient's intimacy).

There is no additional regulation about medical data in the Spanish legislation, but organizations such as *Insalud*[6] have drafted internal regulations about the allowable use of medical data by their employees [101].

A.5 Security Measures for Medical Information Systems

As mentioned before, medical information data are, for the Spanish and European legislations, personal data. For this reason, all security measures from Royal Decree 994/1999 [182] are applicable to systems which process medical data, such as case-histories, histology, etc. This decree already includes all measures mentioned in the Recommendation R(97)5.

The regulations of the Royal Decree establish the minimum security measures of information physical media, with three levels of security, depending on the type of information:

- *High level*: applies to archives containing data about ideology, religion, beliefs, racial origins, health or sexual life, or data collected for police investigation purposes without the consent of the affected persons.

- *Intermediate level*: applies to archives containing data about administrative or penal infringements, Public Finances, financial services, or capital resource solvency and credit.

- *Basic level*: applies to the rest of personal data archives.

Therefore, medical data are typified at the highest security level, and hence, all suitable and adequate organizational and technical measures must be taken, in order to accomplish the following principles:

---

[6]The Spanish National Health Institute.

1. *User identification*: it is mandatory to introduce identification[7] and authentication[8] mechanisms for the persons and institutions with authorization to access and/or use all or part of the data.

2. *Facility access control*: no unauthorized person must be able to access to facilities where personal data are stored or processed.

3. *Data media control*: reading, copying, altering or taking away the data media must be prevented to unauthorized persons.

4. *Memory and telematic transmissions*: unauthorized inputs, queries, modifications or deletions of the data while they are stored in the computer memory of the information system, as well as while the data are sent through the network from a computer to another, must be avoided.

5. *Usage control*: data must be protected against any kind of unauthorized processing, including the unauthorized alteration and communication of such data.

6. *System design*: as a general rule, the design of data structures, procedures, and allowed selective accesses must be such that it allows the separation of a) identifiers and data related to person identity, b) administrative data, c) medical data, d) social data, and e) genetic data.

7. *Data loss protection*: all organizational and technical measures must be taken in order to protect the data against accidental or illegal destruction, and against accidental loss.

8. *Data recovery*: the information system must have a correct backup system, and procedures for recovery in case of partial or full data destruction or loss must be defined beforehand.

9. *Access and data input logging*: the system must guarantee that it will be possible to establish and verify *a posteriori* when and who accessed the system, and which information has been entered.

10. *Incidence log*: it is mandatory to maintain a log of all occurred incidences, be they: unauthorized access attempts, successful or not; system failures which gave rise to the need of restoring information from a backup, etc.

A.6　Desirable Characteristics of a Medical Information System

Current technology allows the demand of high levels of security, privacy, and confidentiality, high enough to allow a generalized and safe usage of patient-related medical information in electronic format. This is also possible although the intended system is local network-distributed or is accessible from the Internet.

---

[7]Procedure to acknowledge the user identity.

[8]Procedure to check the user identity.

Such a system must provide the following functionalities:

1. *Availability*: the system must guarantee that the available information is not only accurate but also up to date. It must also guarantee that this information will reach the user who requests it in perfect conditions and, at the same time, every request must be supervised to confirm its validity.

2. *Access control*: The system must allow total or partial access to the data according to the authorization level of the subject or institution that tries to access the information.

3. *Perimeter identification*: The system must know the physical and logical limits of the kind of accesses it must allow, and it will avoid accesses out of those limits, or the suspicious ones.

In the case of the access control, as we have stated before, this control should be done at different levels:

1. *Access to facilities*: the admission to every facility which holds a part of system hardware must be controlled, be it by means of electronic access control systems, alarms, hired security service, etc.

2. *Telematic access*: network security measures (such as *firewalls*) must be taken in the existing networks to protect those computers telematically connected with the outside world from any unauthorized telematic access.

3. *User identification and authentication*: the minimum authentication level can be achieved by access keys (usually an user's name or code followed by a keyword or *password*). But nowadays there already exist applicable measures in order to increase the identification and authentication level, not only in direct access (*in situ*) to facilities and computers (magnetic strip cards, chip cards, voice recognizers, retina recognizers, etc.) but also in telematic access (digital certificates and signatures).

4. *Access to data in computer's memory*: the control here lies in having an operating system with a good memory management on each computer (especially if it is shared by several users).

5. *Access to stored data*: It is necessary to prevent anyone from obtaining information by a mere copy of data just in the raw format in which they are stored in the physical media. So, besides the access control, it is advisable (although not essential) to save the data in a way that keeps them from being "read" directly. The most usual technique is data *encryption*. It is possible to choose from: a) full data encryption, so the data are completely unreadable if the decryption key is not available, or b) encryption of the part of data allowing individual identification (names, identity cards, etc.) and the part of the data which stores the access keys (keywords or *passwords*), in a way that the only readable data is an amount of anonymous information, and it is not possible to relate it to any identifiable person.

6. *Access to data during telematic communication*: if the system is distributed among several computers and they pass on medical data between them telematically, it is essential to prevent these data from being read by a non-authorized individual who could "capture" them while they travel through the network. This is specially important if uncontrolled telematic networks are used, such as the Internet; in this type of networks we cannot avoid the accidental or intentional capture of data transmissions. In such cases, the most used technique is data encryption, but there exist several variants: private key encryption, combined public and private key encryption, encryption using digital certificates and digital signatures,[9] etc.

A.7 The New Users of Medical Data

The large amount of information generated in a medical action and the need of relating this information with another available information from the same patient, or integrate it in databases that will allow to extract conclusions from a larger population, have led to the appearance of new actors. These organizations supply new services and products to the medical sector, whose existence and success greatly depends on the collection and use of the patient-related information. For instance, we can mention medical equipment suppliers, pharmaceutical companies, reference laboratories, information system suppliers, insurance companies, etc.

Moreover, the alliance between some of these organizations allows to gather data in a larger scale and this dramatically increases its value, not only from a computer science standpoint but also from the economic one. And it is because of this characteristic that they become the possible target of an attack [10] from those who are looking for information.

A.8 Summary

Medical data privacy is an individual's right which can never be given up so it must be taken into account to operate on medical data in electronic format. On the other hand, this operation represents itself an advance which allows to offer better services to the patients. Medical case-histories in electronic format provide a more exact image of the patient, which makes the practitioner's work easier. Hence, it is very important for the benefit of users themselves to know the advantages these formats offer, the security conditions of technology and law, and the limitations of both.

In this appendix we have reviewed some of the requirements to be met by any medical application which manages patient records. We described which are the functional requirements that medical community may require in a software application and also the

---

[9]Digital signatures are ruled by Royal Decree-Law 14/1999 [181].

[10]In the computer security world, an attack is defined as a unauthorized access to a computer and to the data it stores.

imposed requirements in the case of Spanish law. Then we have described some of the security measures to be taken order to meet those requirements.

Protection mechanisms of medical information must also deal with the need for information on the part of the service suppliers, maintaining a proper balance with the confidentiality due to the service recipients.

It is clear that the organizations themselves are who must authorize and identify the users -individuals or institutions- that can access to the databases containing information about their patients, and they also have to control the circumstances in which this access is allowed.

Appendix B

The UCTx System

In this appendix we present a system called *UCTx* designed to model and automate some of the tasks performed by a Transplant Coordination Unit (UCTx) inside a Hospital. The aim of this work is to show how a multi-agent approach allows us to describe and implement the model, and how *UCTx* is capable of interacting with *CARREL* , an Agent Mediated Institution for the Exchange of Human Tissues among Hospitals for Transplantation in order to meet its own goals, acting as the representative of the hospital in the negotiation. As an example we introduce the use of this Agency in the case of Cornea Transplantation.

B.1 Creating a Multi-Agent System for a Hospital's Transplant Unit

Our implementation of an *UCTx* has to reflect the infrastructure and staff of a real Transplant Coordination Unit which allows the successful conclusion of an organ or tissue[1] procurement and extraction process for transplantation [127]. But it also includes new procedures designed to support the distributed computational system. In addition, it deals with the management of the requests for pieces made by the surgeons in order to transplant them into recipients and, it has to follow the sanitary and economic policies that the Hospital dictates.

The *UCTx* implementation has been developed in JADE[2] [16], a framework to create agent platforms that implements all the agent management and communication protocols, leaving to programmers the definition of the agents' internal logic and their behaviors. Each hospital owns a JADE platform (a *UCTx* platform) that communicates with *CARREL*.

---

[1]From now on we will use the word *pieces* to designate organs or tissues or bones.

[2]JADE stands for Java Agent DEvelopment framework, a Java-based environment that follows the FIPA specifications for agent management, ACL message structure, communicative acts and interaction protocols.

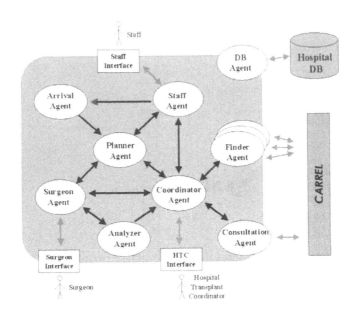

Figure B.1: The Transplant Coordination Unit's agent architecture

A UCTx is modeled as an Agency (see figure B.1) that has several agents, each one competent in a specific task and with its own role and goals to achieve [173]. The Agency is composed of a *Coordinator Agent, Surgeon, Analyser, Finder, Planner, Consultation* and *Arrival* agents.

We can identify the following fundamental services that *UCTx* should provide to those agents:

1. Information confidentiality

2. Information integrity

3. Dynamic accessibility

The first two items are related to data security. As each (software or non-software) agent has a certain role in the system, we are choosing a role-based access mechanism for data confidentiality control [119] (that is, different roles have different views of the data). The data integrity is delegated to the data base controller. The last item is related with the challenge to design and develop clinical systems which are intuitive to use and adequately expressive to satisfy a number of broad requirements: clinical generality, medical expressiveness, operational flexibility, soundness and safety, verifiability and support for reusability. There exist some efforts in this line as GRAIL [184] or PRO*forma* [77] that we will follow.

The *Staff Agent* is an agent to assist and represent any member of the personnel involved in transplants, with the exception of surgeons and the Hospital Transplant Coordinator, which are more specialized. The *Staff Agent* helps the people to plan their daily tasks, acting as their electronic diary and as a message receiver for communications from the Hospital Transplant Coordinator through the *Coordinator Agent*. The agent implementing this role also gives, for each person, access to the clinical information of patients in their care.

The *Surgeon Agent*, specific for each kind of piece, is responsible of communicating with the surgeons through the *Surgeon Interface* and it collects and formalizes the requests for pieces for transplant. Each request has to include the relevant information about the patient, the required piece, optional medical and economic restrictions and the *Selection Function*. The *Surgeon Agent* is able to specialize the *Selection Function* for a given patient using the relevant information coming from the surgeon and the patient's data (see Section B.2). This function provides a way to evaluate each piece for a given offer in the *Institution*[3] *for Human Tissue Exchange*. The *Institution*, in this case *CARREL* assigns the pieces to the different competitor finder agents, one for each hospital and piece, maximizing the satisfaction degree of each request.

The *Surgeon Agent* sends the request to the *Analyser Agent* (see message H1 in table B.1). This agent, specific for each kind of piece, will check if the information was properly introduced, that is, if all the characteristics needed were entered, and if the values are consistent following a given protocol. If there is some data missing, it informs the *Surgeon Agent* who will ask the surgeon to enter or to modify the data in order to validate it. When the *Analyser* has all the information required, it sends the request to the *Coordinator Agent* (see message H3 in table B.3).

The *Coordinator Agent* is responsible for the distribution and coordination of the different tasks that make up the whole process. The *Hospital Transplant Coordinator* in person can communicate with this Agent through the *Coordinator Interface*, and control the behavior of the agent. The *Coordinator Agent* has the following tasks:

1. To coordinate all the agents involved in each transaction.

2. To check all the surgeons' requests for tissues to ensure they meet the hospital protocols.

3. To create, for each checked request, a new *Finder Agent* (see message H4 in table B.4), that will be the one going to the *Institution* to look for the desired piece. The *Coordinator Agent* will give to the *Finder Agent* the request inside an electronic sealed envelope.

4. To keep records of all the piece requests made by the hospital in representation of its surgeons.

5. To give feed-back to the *Institution* when the piece arrives, after transplantation and three weeks after the operation, or in the case of any fatality.

---

[3]Here we follow North's definition [151] of an *Institution* as a collection of artificial constrains that shape human interaction (see §1.3).

Msg♯	Predicate	Parameters
H1	piece_request	id_request, id_piece, piece_params, info_recipient, preferences, urgency_level

Table B.1: Messages Surgeon Agent → Analyser Agent

Msg♯	Predicate	Parameters
H2	analysis_result	id_request, ok | (error, reason)

Table B.2: Messages Analyser Agent → Surgeon Agent

Msg♯	Predicate	Parameters
H3	valid_piece_request	id_request, id_piece, piece_params, info_recipient, preferences, urgency_level

Table B.3: Messages Analyser Agent → Coordinator Agent

Msg♯	Predicate	Parameters
H4	init_request	id_request, id_piece, piece_params, info_recipient, preferences, urgency_level
H5	stop_request	id_request, reason
H6	piece_problem_answer	id_request, new_preferences | accept | refuse

Table B.4: Messages Coordinator Agent → Finder Agent

Msg♯	Predicate	Parameters
H7	piece_assignation	id_request, (piece_info, delivery_plan) | none
H8	piece_problem_question	id_request, info_problem

Table B.5: Messages Finder Agent → Coordinator Agent

6. To give feed-back of all the information about tissue requests (surgeon who made the request, request status and so on) the *Hospital Transplant Coordinator* wants to know in real-time or also by a query in the *Coordinator Interface*. All the feed-back is displayed through the mentioned *Coordinator Interface*, which also allows the *Hospital Transplant Coordinator* to set the amount of real-time information he wants to be displayed.

7. Certify that any access to the hospital's medical data is performed by an authorized agent.

8. To create, for each query requesting information about tissues, a new *Consultation Agent* (see message H13 in table B.7), that will be the one going to the *Institution* to ask for the desired information.

The *Finder Agent* is provided with a sealed envelope with all the information required (e.g., hospital information, patient's data, selection function, etc.). When a *Finder*

Agent returns from *CARREL*, it communicates the result of the negotiation to the *Coordinator Agent* (see message H7 in table B.5). If a piece is found, the *Coordinator Agent*, as part of its agent coordination task, passes the delivery plan proposed by the Institution and the relevant information about the request to the *Planning Agent* (see message H17 in table B.11), which will make up a logistic plan for the reception and transplantation. This information has to arrive to the surgeon that will perform the transplant, too. If no piece was found, the *Coordinator* asks the *Surgeon Agent* to inform the surgeon of such failure. The surgeon can then revise and resubmit the request, or perhaps this can provoke an impasse situation that can only be resolved by the *Hospital Transplant Coordinator* in person. The *Surgeon* and the *Coordinator* agents can stop the process of a request at any moment, if needed.

The *Planner Agent* is responsible for creating the transplant plan, that is, finding a surgery room to match the arrival time of the piece and the surgeon's available schedule. The *Planner Agent* can send several proposals to the surgeon through the *Surgeon Agent* (see message H15 in table B.9). When the surgeon agrees with one of them, the *Planner* will carry out the transplant plan and will also send a message to the *Coordinator Agent*. Otherwise, the *Surgeon Agent* can re-use a given proposal or to create its own proposal and then send it to the *Planner*. If the proposal cannot be carried out, the *Planner* can ask for help to the *Hospital Transplant Coordinator* in person and/or notify the problem to the surgeon. The special characteristics of planning in the Medical Domain are discussed by Miksch [140]. We are following Decker's approach for task planning [52].

The *DB Agent* verifies that each access made by agents in *UCTx* to the clinical data in the Hospital Database is authorized, following a role-based access policy [119].

The *Arrival Agent* is responsible for updating the *Planner Agent* about events that can change the delivery plan (see messages H23 and H24 in table B.13), events that can occur while the transportation of the tissue is made from the Tissue Bank to the Hospital. Then the *Planner* will be able to modify the delivery or the transplant plans dynamically. That is of special interest when there are problems with the delivery (for example, when the hospital receives the wrong piece or the received piece is defective).

Finally the *Consultation Agent*, which is the interface with the Institution's database, processes the different types of queries sent by the *Surgeon Agent*, the *Planner Agent* or the *Coordinator Agent*. This is done, as Finder Agents do, by going to the *CARREL* institution. As the *CARREL* consultation procedures also follow the role-based access model, different levels of privilege are defined by the *CARREL* institution to restrict the access to its database, so, for instance, queries created in the *Coordinator Agent* have a higher privilege level than the ones created inside the Surgeon Agent or the *Planner Agent*, and having a higher privilege level means having access to a wider amount of information in the Institution's database. The *Consultation Agent* has to cope with this privileged levels and the possible rejection of all or part of the information requested.

The information required by the *Finder Agent* to look for a piece in the *CARREL* is packed in an electronic *Sealed Envelope*. This envelope, is created by the *Coordinator Agent* after it has received a valid piece request from the *Surgeon Agent*. The content of the envelope (*urgency level, Hospital identification, coordinator's electronic signature, piece information* and *selection function*) follows *CARREL* specifications, and is described in

Msg♯	Predicate	Parameters
H9	piece_problem_answer	id_request, new_preferences | accept | refuse
H10	query_request	id_query, info_request
H11	transplant_info	id_request, transplant_data
H12	stop_request	id_request, reason

Table B.6: Messages Surgeon Agent → Coordinator Agent

Msg♯	Predicate	Parameters
H13	query_request	id_query, info_request

Table B.7: Messages Coordinator Agent → Consultation Agent

Msg♯	Predicate	Parameters
H14	query_result	id_query, info_result

Table B.8: Messages Consultation Agent → Coordinator Agent

Msg♯	Predicate	Parameters
H15	ask_surgeon_availability	id_request, {time_ini, time_end}*, id_piece

Table B.9: Messages Planner Agent → Surgeon Agent

Msg♯	Predicate	Parameters
H16	tell_surgeon_availability	id_request, time_ini, time_end*, id_piece

Table B.10: Messages Surgeon Agent → Planner Agent

Section 2.4 (tables 2.2 and 2.3).

The data needed by the *Coordinator Agent* in order to create this envelope comes from different sources that may be geographically distributed. The hospital identification and the electronic signature are issued by *CARREL* and only known by the *Coordinator Agent*. All the medical data about the piece and recipient are provided by the *Surgeon Agent* and the *Analyser Agent*.

Once the envelope is created, it is delivered to the *Finder Agent*, which will send it to the *Institution* to look for the piece.

The envelope is an important piece of information that helps to protect the recipient's medical data and to improve and speed the assignment process. The envelope contains almost all the necessary information for the *Finder Agent* to perform its tasks.

B.2 An Example: The Cornea Transplantation

We will use the Cornea Transplantation process to illustrate our ideas. Unlike most tissues in the body, the cornea contains no blood vessels. The cornea must remain transparent to

Msg#	Predicate	Parameters
H17	planning_request	id_request, delivery_plan, info_request
H18	query_request	id_request, info_request
H19	query_result	id_request, info_result

Table B.11: Messages Coordinator Agent → Planner Agent

Msg#	Predicate	Parameters
H20	planning_answer	id_request, delivery_plan, transplant_plant
H21	query_request	id_request, info_request
H22	query_result	id_request, info_result

Table B.12: Messages Planner Agent → Coordinator Agent

Msg#	Predicate	Parameters
H23	piece_reception	id_request, id_piece, state
H24	transport_problem	id_request, info_problem

Table B.13: Messages Arrival Agent → Planner Agent

refract light properly, and the presence of even the tiniest blood vessels can interfere with this process. To see properly, all layers of the cornea must be free of any cloudy or opaque areas. The cornea is as smooth and clear as glass and it helps the eye in two ways:

- It helps to shield the rest of the eye from germs, dust, and other harmful matter. The cornea shares this protective task with the eyelids, the eye socket, tears, and the sclera (see figure B.2).

- The cornea acts as the eye's outermost lens. It functions like a window that controls and focuses the entry of light into the eye. The cornea contributes between 65-75 % of the eye's total focusing power.

A corneal transplant involves replacing a diseased or scarred cornea with a new one. In corneal transplant surgery, the surgeon removes the central portion of the cloudy cornea and replaces it with a clear cornea (see figure B.2), usually donated through a Tissue Bank (TB). A trephine is used to remove the damaged cornea. The surgeon places the new cornea in the opening and sews around it to connect it. The need to automate part of the transplant coordination procedures can be seen if we take a look to the statistics show: in the United States over 40,000 cornea transplants are performed each year (in 1996 there were 46300, that is, 178 pmp[4])[152], [24], [147]; in Catalonia the number of transplants is increasing: in 1999, 845 cornea transplants were performed, that is, 141 pmp, and in 2000 there were 929 transplants, that is, 152 pmp [155]. So the number of transplants is growing so much that Human Transplant Coordinators are beginning to be overwhelmed by the requests.

[4]pmp stands for per million of population.

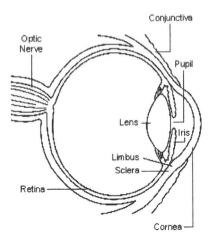

Figure B.2: Diagram of a human eye

The chances of success of this operation have also risen dramatically because of technological advances in the procurement, examination, preservation and implantation procedures and the improvement of the post-implant treatments. For instance, a study supported by the National Eye Institute (NEI) suggests that matching the blood type, but not tissue type, of the recipient with that of the cornea donor may improve the success rate of corneal transplants in people at high risk for graft failure. In US, approximately 20 % of corneal transplant patients–between 6000-8000 a year–reject their corneal grafts. The NEI-supported study, called the Collaborative Corneal Transplantation Study [42], also concluded that intensive steroid treatment after transplant surgery improves the chances for a successful transplant.

So the aim of the UCTx system is not only to automate an important part of the procedures but also allow faster and more exhaustive ways to find the proper cornea[5] for a given recipient, and also store data to monitor the whole process and to make further analysis of the results to detect any critical point in the procedures and then make the suitable improvements.

The Selection Function

The *Selection Function* is a private piece of knowledge given by surgeons to guide the search for suitable corneas made by the *Finder Agent*. This function is composed of a set of rules, each one a constraint the selected piece (e.g., a cornea) has to satisfy. Some of these rules belong to the policy of the whole transplant unit of the hospital, and the rest of the rules are introduced by the surgeon, who can set the constraints needed for a given recipient.

---

[5]We should note here that cornea transplants are only an example of use of the *UCTx* system, as it is obvious that a similar analysis can be done to use *UCTx* with other kind of tissues such as skin or bones.

The rules comprised in such a selection function are the following:

- predicates about the piece: age of the donor, density of Epithelial cells in the cornea.
- predicates about the Tissues Bank: constraints about the Tissue Bank.
- predicates about the cost of the cornea: maximum cost for the selected cornea. An example of such predicate is $(< Cost\ 600euros)$.

The information required to create the selection function comes in part from the *Surgeon Agent*, reflecting the surgeon's preferences for the piece to get, and in part from the *Coordinator Agent*, reflecting the coordinator or hospital's preferences (such as costs, preferred Tissue Banks, etc.).

As an example let us describe an imaginary recipient r with the predicate rule P_r as:

$$P_r = \{ \quad (= Age_r\ Young) \wedge (= Blood\_Type_r\ A) \\ \wedge (= Sex_r\ Male) \wedge ...\} \tag{B.2.1}$$

and he needs a cornea for transplantation. The *UCTx* will prepare an envelope with the petition that will include the encrypted recipient information shown in B.2.1 and the *Selection Function* shown in B.2.2. This will be carried by the *Finder Agent* to the *Institution*.

$$(= Age_d\ Young) \wedge (= TB\ HSCSP) \\ \wedge (> EC_d\ 2000/mm^2) \tag{B.2.2}$$

where Age_d stands for the donor's age, TB stands for Tissue Bank (in the example, HSCSP is the bank of the Hospital de la Santa Creu i Sant Pau) and EC_d stands for the Endothelial Cells density in the donor's cornea. The definition of concepts like $Young$ is settled by the Transplant Coordinator or by the Surgeons using an Ontology that is unique for each Hospital. Of course, each institution could specialize and customize its Selection Functions to fit with their policies. For example, in those countries where donors are in limited supply, to ask for $(> EC_d\ 2000/mm^2)$ and $(= Age_d\ Young)$ may exclude all available corneas in a TB so the *UCTx* should lower these constrains to some other more *acceptable*.

If we modify the recipient characteristics in B.2.1 by doing $(= Age_r\ Old)$ then we can have the following Selection function:

$$(= TB\ HSCSP) \wedge (> EC_d\ 2000/mm^2) \tag{B.2.3}$$

which in turn is more *flexible* than B.2.2.

As each kind of transplant procedure (Cornea Transplant, Lamelar Transplant, Keratoconus Transplant...) has different needs, there will be different rules for each one, and this means different selection functions. If we add to B.2.1 the following information $(= Transplant\ K)$, where K stand for Keratoconus, then B.2.2 will change to:

$$(= Age_d\ Young) \wedge (= TB\ HSCSP) \\ \wedge (> EC_d\ 2800/mm^2) \tag{B.2.4}$$

or even to

$$(= Age_d\ Young) \wedge (=\ TB\ HSCSP)$$
$$\wedge (>\ EC_d\ 2800/mm^2)$$
$$\wedge (=\ Erosion\_Ep_d\ False) \tag{B.2.5}$$

where $Erosion\_Ep_d$ expresses whether there is erosion in the donor's Epithelial layer of the cornea. It is possible to specialize B.2.2, B.2.3, B.2.4 and B.2.5 by adding the following predicates

$$(=\ HLA_d\ DR) \wedge (=\ Blood\_Type_d\ AB0)$$

The HLA predicate will measure the histocompatibility between the Donor and the Recipient, although this is only important when a potential recipient had suffered from previous graft rejections.

Surgeons or the Hospital Transplant Coordinator can introduce other constraint rules about the cornea, such as the time it has been in preservation at the Tissues Bank:

$$(=\ Age_r\ Young) \wedge (<\ Hours\_In\_TB\ 72) \tag{B.2.6}$$

as some surgeons think that corneas with more than 3 days (72 hours) inside the TB are not good choices for a young recipient.

The surgeons can easily create their own rules to build their own selection functions by means of a rule editor in the *Surgeon Interface*. With this editor a surgeon can compose a rule, and then associate a weight to each rule. This interface keeps all rules to be re-used in future requests. This rule base will constitute the *memory* of each service and should be exploited to learn better strategies to create new rules. The weights associated to rules allow the *Finder Agent* to know which of the rules are more important than others while it is searching for a cornea and, as it was introduced in Section B.1, the weights allow to qualify each piece and create a partial order among them.

B.3   Summary

The social pressure and the need to improve the quality of health care has lead to a strong demand for clinical protocols and, lately, computer systems supporting both their creation and the execution. Organ and Tissue transplants in general, and corneal transplants in particular often are the best technique for the treatment of some major health problems that can affect the quality of life of an important part of the population. So improving the success of such techniques is very important.

In this appendix we present the *UCTx* agency, a Multi-Agent System that models the interaction of the different actors of a Transplant Coordination Unit inside a Hospital. The *UCTx* system interacts with an Agent Mediated Institution for Human Tissue Exchange (see *CARREL* in Chapter 2), and this collaboration ensures that the process meets the protocols and the rules established by national transplant organizations and hospitals.

UCTx speeds up the process by its automation, which can reduce the time since the extraction of the tissue to its implant in the recipient, increasing the quality of the piece implanted. That is of special relevance in the cornea's transplantation, as corneas are perishable. In other cases, the use of the systems will help in finding the *best* stored item for a given patient.

The *UCTx* platform can be very helpful in a Transplant Coordination Unit as it can aid in some of the daily management issues such as coordination of surgeons or planning of operations and even automate some tedious tasks such as looking for an available surgery room or more specialized as looking for a proper tissue. It will help to maintain the medical data integrity and confidentiality by supervising every access to data and keeping track of any data demand, see for example [145]. This is coherent with the new Spanish law on personnel data protection [118] [182] and meets also the European Directive 97/66/CE on Data Protection [62].

UCTx can assist the surgeons while they build their requests for pieces (see Section B.2), manage the requests and inform the Hospital Transplant Coordinator of any important event that occurs.

On the other hand, as the system asks for a complete clinical evaluation of each piece, it can decrease the cost of transplants by reducing the number of unsuitable transplants and furthermore offering a higher security level in reducing the chances of possible infections.

One additional advantage of using such a system is that all the information that *UCTx* gathers about organs, tissues, bones and recipients is organized using an electronic format that can be analysed later to obtain through data mining or machine learning procedures. Discovery of new knowledge by mining medical databases is crucial in order to make an effective use of stored data to generate new useful knowledge about the transplant process, knowledge which can lead to its improvement, from the enhancement of the tissue selection functions to the extraction and implantation procedures.

Bibliography

[1] A. Abdul-Rahman and S. Hailes. Supporting trust in virtual communities. In *Proceedings of the Hawaii International Conference on System Sciences*, 2000.

[2] V. Akman and F.N. Alpaslan. Strawson on intended meaning and context. In *Modelling and Using Context: Proceedings of the Second International and Interdisciplinary Conference, CONTEXT'99, Trento, Italy*, Lecture Notes on Artificial Intelligence 1688, pages 1–14. Springer-Verlag, September 1999.

[3] A. Aldea, B. López, A. Moreno, D. Riaño, and A. Valls. A multi-agent system for organ transplant co-ordination. In Barahona Quaglini and Andreassen, editors, *Proceedings of the 8th. European Conference on Artificial Intelligence in Medicine, Portugal, 2001.*, Lecture Notes in Artificial Intelligence 2101: Artificial Intelligence in Medicine, pages 413–416, 2001.

[4] T. Alsinet, R. Béjar, C. Fernàndez, and F. Manyà. A multi agent system architecture for monitoring medical protocols. In C. Sierra, M. Gini, and J. Rosenschein, editors, *Proceedings of the Fourth International Conference on Autonomous Agents*, pages 499–505. ACM-AAAI, ACM Press., 2000.

[5] The Administrative Normative Information Transaction Agents Project (ANITA). http://rint.rechten.rug.nl/onderzoek/anita/anita.html.

[6] Journal on Artificial Intelligence in Medicine. http://www.harcourt-international.com/journals/aiim/.

[7] K.D. Ashley. *Modelling Legal Argument: Reasoning with Cases and Hypotheticals*. MIT Press, Cambridge, MA, 1990.

[8] I. Asimov. *The Gods Themselves*. Pimlico Books Limited, 1972.

[9] R. Axelrod. *The Evolution of Cooperation*. New York: Basic Books, 1984.

[10] R. Axelrod and D. Dion. The further evolution of cooperation. *Science*, (242):1385–1390, 1988.

[11] W. Balzer. A Basic Model for Social Institutions. *Journal of Mathematical Sociology*, 16(1):1–29, 1990.

[12] W. Balzer and R. Tuomela. Social institutions, norms and practices. In C. Dellarocas and R. Conte, editors, *Workshop on Norms and Institutions in Multi-Agent Systems*, pages 2–12. ACM-AAAI, ACM Press, 2000.

[13] M.D. Beer, T. Bench-Capon, and A. Sixsmith. Using agents to deliver effective integrated community care. In V. Shankararaman, editor, *Workshop on Autonomous Agents in Health Care*, pages 35–45. ACM-AAAI, ACM Press, 2000.

[14] J. Bell. Pragmatic reasoning: Inferring contexts. In *Modelling and Using Context: Proceedings of the Second International and Interdisciplinary Conference, CONTEXT'99, Trento, Italy*, Lecture Notes on Artificial Intelligence 1688, pages 42–53. Springer-Verlag, September 1999.

[15] J. Bell. Pragmatic reasoning: pragmatic semantics and semantic pragmatics. In *Modelling and Using Context: Proceedings of the Third International and Interdisciplinary Conference, CONTEXT 2001, Dundee, UK*, Lecture Notes on Artificial Intelligence 2116, pages 45–58. Springer-Verlag, July 2001.

[16] F. Bellifemine, A. Poggi, and G. Rimassa. JADE : A FIPA compliant agent framework. In *Proceedings of the International Conference on the Practical Application of Intelligent Agents and Multi-Agent Systems (PAAM'1999)*, pages 97–108, 1999.

[17] C. Berthouzoz. A model of context adapted to domain-independent machine translation. In *Modelling and Using Context: Proceedings of the Second International and Interdisciplinary Conference, CONTEXT'99, Trento, Italy*, Lecture Notes on Artificial Intelligence 1688, pages 54–66. Springer-Verlag, September 1999.

[18] C. Bianchi. Three forms of contextual dependence. In *Modelling and Using Context: Proceedings of the Second International and Interdisciplinary Conference, CONTEXT'99, Trento, Italy*, Lecture Notes on Artificial Intelligence 1688, pages 67–76. Springer-Verlag, September 1999.

[19] F.C. Billari. Social norms and life course events: a topic for simulation? In C. Dellarocas and R. Conte, editors, *Workshop on Norms and Institutions in Multi-Agent Systems*, pages 13–14. ACM-AAAI, ACM Press, 2000.

[20] E. Black, J. Vázquez-Salceda, J. Fox, and U. Cortés. Formalising and Simulating a Medical Institution Using PROforma Agents. Technical report, Advanced Computation Laboratory, Cancer Research UK, London UK, 2003.

[21] G. Boella and L. Lesmo. Deliberative normative agents. In C. Dellarocas and R. Conte, editors, *Workshop on Norms and Institutions in Multi-Agent Systems*, pages 15–25. ACM-AAAI, ACM Press, 2000.

[22] A. Boer, R. Hoekstra, and R. Winkels. The CLIME Ontology. In *In Proceedings of the Second International Workshop on Legal Ontologies*, pages 37–47. University of Amsterdam, 2001.

[23] P. Bouquet and L. Serafini. Two formalizations of context: a comparison. In *Modelling and Using Context: Proceedings of the Third International and Interdisci-*

plinary Conference, CONTEXT 2001, Dundee, UK, Lecture Notes on Artificial Intelligence 2116, pages 87–102. Springer-Verlag, July 2001.

[24] F.S. Brightbill, editor. *Corneal surgery: Theory, techniques and tissue.* Mosby Inc., 1999.

[25] S. Buvač and I.A. Mason. Propositional logic of context. In R. Fikes and W. Lehnert, editors, *Proceedings of the 11th National Conference on Artificial Intelligence*, pages 412–419. AAAI Press, 1993.

[26] S. Buvač and J. McCarthy. Combining planning contexts. In Austin Tate, editor, *Advanced Planning Technology–Technological Achievements of the ARPA/Rome Laboratory Planning Initiative.* AAAI Press, 1996.

[27] D. Cabanillas, S. Willmott, and U. Cortés. Threats and security safeguards in a multi-agent system for medical applications. Technical Report LSI-02-76-R, Departament de Llenguatges i Sistemes Informàtics, 2002.

[28] L.M. Camarinha-Matos and H. Afsarmanesh. Virtual communities and elderly support. In V.V. Kluev, C.E. D'Attellis, and N. E. Mastorakis, editors, *Proceedings of MIV'01 in Advances in Automation, Multimedia and Video Systems, and Modern Computer Science*, pages 279–284. WSES press, 2001.

[29] L.M. Camarinha-Matos and W. Viera. Using multiagent systems and the internet in care services for the ageing society. In L.M. Camarinha-Matos, H. Afsarmanesh, and V. Marík, editors, *Intelligent Systems for Manufacturing: Multi-Agent Systems and Virtual Organizations, Proceedings of the BASYS - 3rd IEEE/IFIP International Conference on Information Technology for Balanced Automation Systems in Manufacturing, Prague, Czech Republic*, volume 130 of *IFIP Conference Proceedings*, pages 33–48. Kluwer, 1998.

[30] L. Cardelli. Mobility and security. In Friedrich L. Bauer, editor, *Proceedings of the NATO Advanced Study Institute on Foundations of Secure Computation*, number ISBN 1 58603 015 9 in NATO Science Series, pages 3–37. IOS Press, 1999.

[31] K.M. Carley and A. Newell. The nature of the social agent. *Journal of Mathematical Sociology*, 19:221–262, 1994.

[32] J. Carmo and O. Pacheco. Deontic and action logics for collective agency and roles. In R. Demolombe and R. Hilpinen, editors, *Proceedings of the Fifth International Workshop on Deontic Logic in Computer Science (DEON'00),ONERA-DGA, Toulouse*, pages 93–124, 2000.

[33] J. Carmo and O. Pacheco. Logics for Modeling Business and Agents Interaction. In *Proceedings of the International Conference on Advances in Infrastructure for Electronic Business, Science and Education on the Internet (SSGRR 2000), l' Aquilla*, 2000.

[34] C. Castelfranchi. Multi-agent reasoning with belief contexts: the approach and a case study. *Intelligent agents*, 1995.

[35] C. Castelfranchi. Formalising the informal? dynamic social order, bottom-up social control, and spontaneous normative relations. *Nordic Journal of Philosophical Logic*, 5(3), 2000.

[36] C. Castelfranchi, F. Dignum, C. Jonker, and J. Treur. Deliberative Normative Agents: Principles and Architecture. In N.R. Jennings and Y. Lespérance, editors, *Intelligent Agents VI, Agent Theories, Architectures, and Languages (ATAL), 6th International Workshop, ATAL '99, Orlando, Florida, USA, July 15-17, 1999, Proceedings*, volume 1757 of *Lecture Notes in Computer Science*. Springer, 2000.

[37] J. Castro, M. Kolp, and J. Mylopoulos. Towards requirements-driven information systems engineering: the TROPOS project. *Information Systems*, 27:365–389, 2002.

[38] A. Cimatti, E.M. Clarke, F. Giunchiglia, and M. Roveri. NuSMV: A new symbolic model checker. *International Journal on Software Tools for Technology Transfer*, 2(4):410–425, 2000.

[39] O. Cliffe and J. Padget. A framework for checking agent interaction within institutions. In *Proceedings of MOCHART workshop at ECAI'02*, 2002.

[40] P.R. Cohen and H.J. Levesque. Persistence, intention, and commitment. In P. R. Cohen, J. Morgan, and M. E. Pollack, editors, *Intentions in Communication*, pages 33–69. MIT Press, Cambridge, MA, 1990.

[41] P.R. Cohen and H.J. Levesque. Teamwork. *Nous*, 25(4):487–512, 1991.

[42] Collaborative Corneal Transplantation Studies Group. Collaborative corneal transplantation studies (CCTS). Technical report, National Eye Institute, 1999.

[43] J.H. Connolly. Context in the study of human languages and computer programming languages: a comparison. In *Modelling and Using Context: Proceedings of the Third International and Interdisciplinary Conference, CONTEXT 2001, Dundee, UK*, Lecture Notes on Artificial Intelligence 2116, pages 116–128. Springer-Verlag, July 2001.

[44] Constitución española de 1978. Boletín Oficial del Estado 311, 29 de diciembre 1978.

[45] R. Conte and C. Castelfranchi. *Cognitive and Social Action*. UCL Press, 1995.

[46] R. Conte and C. Castelfranchi. Are incentives good enough to achieve(info)social order? In C. Dellarocas and R. Conte, editors, *Workshop on Norms and Institutions in Multi-Agent Systems*, pages 26–40. ACM-AAAI, ACM Press, 2000.

[47] R. Conte and F. Dignum. From social monitoring to normative influence. *Journal of Artificial Societies and Social Simulation*, 4(2), 2001.

[48] U. Cortés, A. López-Navidad, J. Vázquez-Salceda, A. Vázquez, D. Busquets, M. Nicolás, S. Lopes, F. Vázquez, and F. Caballero. Carrel: An agent mediated institution for the exchange of human tissues among hospitals for transplantation. In *3er Congrés Català d'Intel.ligencia Artificial*, pages 15–22. ACIA, 2000.

[49] U. Cortés and J.A. Rodríguez-Aguilar. Trading Agents in Auction-based Tournaments. The EPFL Experience. *INFORMATIK*, 1:39–50, 2000.

[50] U. Cortés, J. Vázquez-Salceda, A. López-Navidad, and F. Caballero. UCTx: A multi-agent system to assist a transplant coordinator unit. *Journal of Applied Intelligence*, 20, 2004.

[51] R. Dawkins. *The Selfish Gene*. Oxford University Press, 2nd edition, 1989.

[52] K.S. Decker. *Task Environment Centered Simulation*, chapter 6, pages 105–128. In Prietula et al. [173], 1st edition, 1998.

[53] C. Dellarocas and M. Klein. Contractual agent societies: Negotiated shared context and social control in open multi-agent systems. In C. Dellarocas and R. Conte, editors, *Workshop on Norms and Institutions in Multi-Agent Systems*, pages 41–52, 2000.

[54] D. Dennet. *Brainstorms*. Harvester Press, 1981.

[55] F. Dignum. Autonomous agents with norms. *Artificial Intelligence and Law*, 7:69–79, 1999.

[56] F. Dignum. Abstract norms and electronic institutions. In G. Lindemann, D. Moldt, M. Paolucci, and B. Yu, editors, *Proceedings of the International Workshop on Regulated Agent-Based Social Systems: Theories and Applications (RASTA '02), Bologna*, volume 318 of *Mitteilung*, pages 93–104, Hamburg, 12 July 2002. Fachbereich Informatik, Universität Hamburg.

[57] F. Dignum, D. Morley, and E.A. Sonenberg. Towards socially sophisticated BDI agents. In *DEXA Workshop*, pages 1134–1140, 2000.

[58] V. Dignum. *A Model for Organizational Interaction: based on Agents, founded in Logic*. SIKS Dissertation Series 2004-1. SIKS, 2004. PhD Thesis.

[59] V. Dignum, J.-J.Ch. Meyer, H. Wiegand, and F. Dignum. An organisational-oriented model for agent societies. In G. Lindemann, D. Moldt, M. Paolucci, and B. Yu, editors, *Proceedings of the International Workshop on Regulated Agent-Based Social Systems: Theories and Applications (RASTA '02), Bologna*, volume 318 of *Mitteilung*, pages 31–50, Hamburg, 12 July 2002. Fachbereich Informatik, Universität Hamburg.

[60] V. Dignum, H. Weigand, and L. Xu. Agent societies: Towards framework-based design. In *Agent-Oriented Software Engineering II: Proceedings of the 2nd Workshop on Agent-Oriented Software Engineering, Autonomous Agents, Montreal, Canada*, Lecture Notes in Computer Science 2222. Springer-Verlag, 2002.

[61] M. d'Inverno and M. Luck. *Understanding Agent Systems*. Springer Verlag, 2001.

[62] Directive 95/46/CE of the European Parliament and of the Council of 24 october 1995 on the protection of individuals with regard to the processing of personal data and of the free movement of such data, October 1995.

[63] The Donor Action Foundation.
      http://www.donoraction.org.

[64] G. Dorais, R.P. Bonasso, D. Kortenkamp, P. Pell, and D. Schreckenghost. Adjustable autonomy for human-centered autonomous systems on mars. *Proceedings of the First International Mars Society Convention, Boulder, CO*, 1998.

[65] S. Egashira and T. Hashimoto. Forming and sharing norms under the assumption of 'fundamental' imperfect information. In C. Dellarocas and R. Conte, editors, *Workshop on Norms and Institutions in Multi-Agent Systems*, pages 53–64, 2000.

[66] The eInstitution Platform.
      http://e-institutor.iiia.csic.es/islander/pub/index.html.

[67] ESCULAPE: Use of computer techniques for tissues matching and analysis as an aid to human transplantation.
      http://dbs.cordis.lu/cordis-cgi/srchidadb?CALLER=EN_CORDIS&
      QZ_WEBSRCH=ESCULAPE.

[68] M. Esteva, J. Padget, and C. Sierra. Formalizing a language for institutions and norms. In J.-J.Ch. Meyer and M. Tambe, editors, *Intelligent Agents VIII*, volume 2333 of *Lecture Notes in Artificial Intelligence*, pages 348–366. Springer Verlag, 2001. ISBN 3-540-43858-0.

[69] Eurotransplant International Foundation.
      http://www.eurotransplant.nl.

[70] N.V. Findler and R. Malyankar. Social structures and the problem of coordination in intelligent agent societies. 2000.

[71] Klaus Fischer, Jörg P. Müller, and Markus Pischel. Unifying control in a layered agent architecture. Technical Report TM-94-05, Deutsches Forschungszentrum für Künstliche Intelligenz GmbH, 1994.

[72] The FishMarket Project.
      http://www.iiia.csic.es/Projects/fishmarket.

[73] D. Fitoussi and M. Tennenholtz. Choosing social laws for multi-agent systems: Minimality and simplicity. *Artificial Intelligence*, 119(1-2):61–101, 2000.

[74] *The Foundation for Intelligent Phisical Agents (FIPA) Specifications.*
      http://www.fipa.org/, 2000.

[75] J. Fox and S. Das. *Safe and Sound.* AAAIPress/MIT Press, 1st edition, 1999.

[76] J. Fox and S. Das. Guardian agents for safety-critical systems. In V. Shankararaman, editor, *Workshop on Autonomous Agents in Health Care*, pages 25–34, 2000.

[77] J. Fox, N. Johns, A. Rahmanzadeh, and R. Thompson. Proforma: A method and a language for specifying clinical guidelines and protocols. In J. Brender, J. P. Christensen, J-R. Scherrer, and P. McNair, editors, *Medical Informatics Europe'96*, pages 516–520. IOS Press, 1996.

[78] D. Gambetta. *Can We Trust Trust?*, pages 213–237. Basil Blackwell, Oxford, 1990.

[79] A. Gangemi, D.M. Pisanelli, and G. Steve. A Formal Ontology Framework to Represent Norm Dynamics. In *Proceedings of the Second International Workshop on Legal Ontologies*. University of Amsterdam, 2001.

[80] P. García, E. Gimenez, L. Godó, and J.A. Rodríguez-Aguilar. Bidding Strategies for Trading Agents in Auction-Based Tournaments. In *Electronic Commerce: selected papers / First International Workshop on Agent-Mediated Electronic Trading (AMET'98)*, Lecture Notes on Artificial Intelligence 1571, pages 151–165. Springer-Verlag, 1999.

[81] J. Gelati, G. Governatori, A. Rotolo, and G. Sartor. Declarative Power, Representation and Mandate. A Formal Analysis. In *Proceedings of the Fifteen Annual International Conference on Legal Knowledge and Information Systems (JURIX 2002)*. Institute of Advanced Legal Studies, London, UK, 2002.

[82] C. Ghidini. Modelling (un)bounded beliefs. In *Modelling and Using Context: Proceedings of the Second International and Interdisciplinary Conference, CONTEXT'99, Trento, Italy*, Lecture Notes on Artificial Intelligence 1688, pages 145–158. Springer-Verlag, September 1999.

[83] C. Ghidini and L. Serafini. A context-based logic for distributes knowledge representation and reasoning. In *Modelling and Using Context: Proceedings of the Second International and Interdisciplinary Conference, CONTEXT'99, Trento, Italy*, Lecture Notes on Artificial Intelligence 1688, pages 159–172. Springer-Verlag, September 1999.

[84] F. Giunchiglia and P. Bouquet. Introduction to contextual reasoning. *Perspectives on Cognitive Science*, 3, 1997.

[85] F. Giunchiglia and C. Ghidini. Local Model Semantics, or Contextual Reasoning = Locality + Compatibility. In *Proceedings of the 6th. International Conference on Principles of Knowledge Representation and Reasoning (KR*98)*. Morgan Kaufmann, 1998.

[86] L. Godó, R. López de Mántaras, J. Puyol-Gruart, and C. Sierra. RENOIR, PNEUMON-IA and TERAP-IA: three medical applications based on fuzzy logic. *Artificial Intelligence in Medicine*, 21(1-3):153–162, 2001.

[87] T.F. Gordon. *The Pleadings Game. An Artificial Intelligence Model of Procedural Justice*. Kluwer Academic Publishers, 1995.

[88] G. Governatori, J. Gelati, A. Rotolo, and G. Sartor. Actions, Institutions, Powers. Preliminary Notes. In G. Lindemann, D. Moldt, M. Paolucci, and B. Yu, editors, *Proceedings of the International Workshop on Regulated Agent-Based Social Systems: Theories and Applications (RASTA'02), Bologna, Italy*, volume 318 of *Mitteilung*, pages 131–147, Hamburg, 12 July 2002. Fachbereich Informatik, Universität Hamburg.

[89] D. Grossi. *Astrattezza Normativa nelle Istituzioni. Proposte per un Modello Logico-Formale.* Università degli Studi di Pisa, April 2003. M.S. Thesis.

[90] D. Grossi and F. Dignum. Abstract and concrete norms in institutions. Technical report, Institute of Information and Computing Sciences, Utrecht University, 2004.

[91] B.J. Grosz and S. Kraus. Collaborative plans for complex group actions. *Artificial Intelligence*, (86):269–358, 1996.

[92] B.J. Grosz and C.L. Sidner. Plans for discourse. In P. R. Cohen, J. Morgan, and M. E. Pollack, editors, *Intentions in Communication*, pages 417–444. MIT Press, Cambridge, MA, 1990.

[93] R.V. Guha. *Contexts: A Formalization and Some Applications.* Standford University, 1991. PhD. Thesis.

[94] J. Hage. Dialectical models in artificial intelligence and law. *Artificial Intelligence and Law*, (8(2/3)):137–172, 2000.

[95] J.C. Hage. *Reasoning with Rules. An Essay on Legal Reasoning and Its Underlying Logic.* Number 27 in Law and Philosophy Library. Kluwer Academic Publishers, 1997.

[96] Health Insurance Portability and Accountability Act of 1996. http://www.hcfa.gov/medicais/hipaa/.

[97] G.J. Holzmann. The Spin model checker. *IEEE Transactions on Software Engineering*, 23(5):279–95, May 1997.

[98] T. Horgan and M. Potrc. *From Epistemicism to Transvaluationism: Perspectives on Vagueness.* Oxford University Press, forthcoming.

[99] Nick Howden, Ralph Rönnquist, Andrew Hodgson, and Andrew Lucas. Jack - summary of an agent infrastructure. In *5th International Conference on Autonomous Agents.* 2001.

[100] J. Huang, N.R. Jennings, and J. Fox. Agent-based approach to healthcare management. *International Journal of Applied Artificial Intelligence*, 9(4):401–420, 1995.

[101] Instrucciones del Insalud sobre seguridad y protección de datos. Circular núm 9/97, 9 de julio 1997.

[102] Internet Medical Terminology Ressources. http://ai.bpa.arizona.edu.

[103] The ISLANDER Graphical Editor. http://e-institutor.iiia.csic.es/islander/islander.html.

[104] N.R. Jennings. Commitments and conventions: The foundation of coordination in multi-agent systems. *The Knowledge Engineering Review*, 8(3):223–250, 1993.

[105] N.R. Jennings. Controlling cooperative problem solving in industrial multi-agent systems using joint intentions. *Artificial Intelligence*, 75(2):195–240, 1995.

[106] A.J.I. Jones and M.J. Sergot. Deontic logic in the representation of law: Towards a methodology. *Artificial Intelligence and Law*, 1(1):45–64, 1992.

[107] A.J.I. Jones and M.J. Sergot. A formal characterisation of institutionalised power. *Journal of the Interest Group in Pure and Applied Logics (IGPL)*, 4(3):429–445, 1996.

[108] H. Kautz and B. Selman. Creating models of real-world communities with ReferralWeb. *Working notes of the Workshop on Recommender Systems, held in conjunction with AAAI-98, Madison, WI*, 1998.

[109] E.G. Kim and N.V. Findler. Toward an automatically generated theory of coordination - empirical explorations. In Frank van Harmelen, editor, *Proceedings of the 15th European Conference on Artificial Intelligence (ECAI 2002)*, pages 3–7. IOS Press, 2002.

[110] K.E. Lange and C.H. Lin. Advanced life support program - requirements definition and design considerations. technical report. Technical Report CTSD-ADV-245, NASA, Lyndon B. Johnson Space Center, Houston, Texas, 1996.

[111] G. Lanzola, L. Gatti, S. Falasconi, and M. Stefanelli. A framework for building cooperative software agents in medical applications. *Artificial Intelligence in Medicine*, 16(3):223–249, 1999.

[112] J.E. Larsson and N. Hayes-Roth. Guardian: an intelligent autonomous agent for medical monitoring and diagnosis. *IEEE Intelligent Systems*, pages 58–64, January-February 1998.

[113] Ley 30/1979, de 27 de octubre, sobre extracción y transplante de órganos. Boletín Oficial del Estado 266, 29 de abril 1986.

[114] Ley 14/1986, de 25 de abril, General de Sanidad. Boletín Oficial del Estado 102, 29 de abril 1986.

[115] Ley 25/1990, de 20 de diciembre, del medicamento. Boletín Oficial del Estado 306, 22 de diciembre 1990.

[116] Ley Orgánica 3/1986, de 14 de abril, de Medidas Especiales en Materia de Salud Pública. Boletín Oficial del Estado 102, 29 de abril 1979.

[117] Ley Orgánica 5/1992, de 29 de octubre, de Regulación del Tratamiento Automatizado de los Datos de carácter personal. Boletín Oficial del Estado 262, 31 de octubre 1992.

[118] Ley Orgánica 15/1999 de protección de datos de carácter personal. Boletín Oficial del Estado 292, 14 de diciembre 1999.

[119] A. Lin. Integrating policy-driven role-based access control with common data secutiry architecture. Technical Report HPL-1999-59, HP Laboratories Bristol, 1999.

[120] J. Lind. *Iterative Software Engineering for Multiagent Systems. The MASSIVE Method*, volume 1994 of *LNAI*. Springer Verlag, 2000.

[121] J. Lind. The Massive Development Method for Multiagent Systems. In Jeffrey Bradshaw and Geoff Arnold, editors, *Proceedings of the 5^{t}h International Conference on the Practical Application of Intelligent Agents and Multi-Agent Technology (PAAM 2000)*, pages 339–354, Manchester, UK, 2000. The Practical Application Company Ltd.

[122] K.E. Lochbaum, B.J. Grosz, and C.L. Sidner. Models of plans to support communication. In *Proceedings of the 8th National Conference on Artificial Intelligence, Boston, USA*, pages 485–490, 1990.

[123] G.-J. C. Lokhorst. Ernst mally's deontik. *Notre Dame Journal of Formal Logic*, 40(2):273–282, 1999.

[124] A. Lomuscio and M.J. Sergot. On Multi-agent Systems Specification via Deontic Logic. In John-Jules Ch. Meyer and Milind Tambe, editors, *Intelligent Agents VIII, 8th International Workshop, ATAL 2001 Seattle, WA, USA, August 1-3, 2001, Revised Papers*, volume 2333 of *Lecture Notes in Computer Science*, pages 86–99. Springer, 2002.

[125] H. Lopes Cardoso and E. Oliveira. Using and Evaluating Adaptive Agents for Electronic Commerce Negotiation. In M.C. Monard and J.S. Sichman, editors, *Advances in Artificial Intelligence, International Joint Conference, 7th Ibero-American Conference on AI, 15th Brazilian Symposium on AI, IBERAMIA-SBIA 2000, Atibaia, SP, Brazil, November 19-22, 2000, Proceedings*, volume 1952 of *Lecture Notes in Computer Science*. Springer, 2000.

[126] A. López-Navidad. Professional characteristics of the transplant coordinator. *Transplantation Proceedings*, (23):1607–1613, 1997.

[127] A. López-Navidad, J. Kulisevsky, and F. Caballero, editors. *El donante de órganos y tejidos: Evaluación y manejo*. Springer-Verlag Ibérica, 1st edition, 1997.

[128] F. López y López and M. Luck. Empowered Situations of Autonomous Agents. In F.J. Garijo, J.C. Riquelme, and M. Toro, editors, *IBERAMIA 2002*, volume 2527 of *LNAI*, pages 585–595, Berlin Heidelberg, 2002. Springer Verlag.

[129] F. López y López and M. Luck. Towards a Model of the Dynamics of Normative Multi-Agent Systems. In G. Lindemann, D. Moldt, M. Paolucci, and B. Yu, editors, *Proceedings of the International Workshop on Regulated Agent-Based Social Systems: Theories and Applications (RASTA '02), Bologna*, volume 318 of *Mitteilung*, pages 175–194, Hamburg, 12 July 2002. Fachbereich Informatik, Universität Hamburg.

[130] F. López y Lopez, M. Luck, and M. d'Inverno. A framework for norm-based inter-agent dependence. In *Proceedings of The Third Mexican International Conference on Computer Science*, pages 31–40. SMCC-INEGI, 2001.

[131] M. Luck, P. McBurney, and C. Priest. *Agent Technology: Enabling Next Generation Computing. A Roadmap for Agent Based Computing*. AgentLink, 2003.

[132] R. Matesanz. Meeting the organ shortage: Current status and strategies for im-provement of organ donation. *Newsletter Transplant*, 4(1):5–17, 1999.

[133] N. Matos and C. Sierra. Evolutionary Computing and Negotiating Agents. In *Electronic Commerce: selected papers / First International Workshop on Agent-Mediated Electronic Trading (AMET'98)*, Lecture Notes on Artificial Intelligence 1571, pages 91–111. Springer-Verlag, 1999.

[134] J. McCarthy. Generality in artificial intelligence. *Communications of the ACM*, 30(12):1030–1035, 1987.

[135] J. McCarthy. *Formalizing Common Sense*. Ablex, Norwoodm New Jersey, 1990.

[136] J. McCarthy and S. Buvač. Formalizing Context (Expanded Notes). In A. Aliseda, R.J. van Glabbeek, and D. Westerståhl, editors, *Computing Natural Language*, vol-ume 81 of *CSLI Lecture Notes*, pages 13–50. Center for the Study of Language and Information, Stanford University, 1998.

[137] P. McNamara and H. Prakken. *Norms Logic and Information Systems*. IOS Press, 1999.

[138] J.-J. Ch. Meyer. A different approach to deontic logic: Deontic logic viewed as a variant of dynamic logic. *Notre Dame Journal of Formal Logic*, 29(1):109–136, 1988.

[139] J.-J. Ch. Meyer and R.J. Wieringa. *Deontic Logic in Computer Science: Normative System Specification*. John Wiley and sons, 1991.

[140] S. Miksch. Plan management in medical domain. *Ai. Communications*, 4:209–235, 1999.

[141] V.O. Mittal, H.A. Yanco, J. Aronis, and R. Simpson, editors. *Assistive Technology and Artificial Intelligence: Applications in Robotics, User Interfaces and Natu-ral Language Processing*, volume 1458 of *Lecture Notes in Artificial Intelligence*. Springer-Verlag, Berlin, 1998.

[142] A. Moreno, A. Valls, and J. Bocio. Management of hospital teams for organ trans-plants using multi-agent systems. In Barahona Quaglini and Andreassen (Eds.), editors, *Proceedings of the 8th. European Conference on Artificial Intelligence in Medicine, Portugal, 2001.*, Lecture Notes in Artificial Intelligence 2101: Artificial Intelligence in Medicine, pages 374–383., 2001.

[143] A. Moreno, A. Valls, and A. Ribes. Finding efficient organ transport routes using multi-agent systems. In *IEEE 3rd International Workshop on Enterprise Network-ing and Computing in Health Care Industry (Healtcom), L'Aquilla, Italy.*, 2001.

[144] Y. Moses and M. Tennenholtz. Artificial social systems. *Computers and AI*, 14(6):533–562, 1995.

[145] National Research Council. Committee on Maintaining Privacy and Security in Health Care Applications of the National Information Infrastructure. *For the Record: Protecting Electronic Health Information*. National Academy Press, 1st edition, 1997.

[146] J.L. Nealon and A. Moreno, editors. *Applications of Software Agent Technology in the Health Care Domain*. Whitestein Series in Software Agent Technologies. Birkhäuser Verlag, Basel, 2003.

[147] The National Eye Institute . http://www.nei.nih.gov.

[148] A. Newell and H. Simon. The knowledge level. *Artificial Intelligence*, 18(1):87–127, 1982.

[149] P. Noriega. *Agent-Mediated Auctions: The Fishmarket Metaphor*. Number 8 in IIIA Monograph Series. Institut d'Investigació en Intel.ligència Artificial (IIIA), 1997. PhD Thesis.

[150] Timothy J. Norman and Chris Reed. Delegation and responsibility. *Lecture Notes in Computer Science*, 1986:136–149, 2001.

[151] D.C. North. *Institutions, Instutional Change and Econonomic Performance*. Cambridge Univ. Press, 1st edition, 1990.

[152] D.M. O'Day and J.M. Khoury. *Donor tissue selection*, pages 869–875. In Brightbill [24], 1999.

[153] A. Omicini. Soda: Societies and infrastructures in the analysis and design of agent-based systems. In P. Ciancarini and M. Wooldridge, editors, *Agent-Oriented Software Engineering*, volume 1957 of *LNAI*, pages 185–193. Springer Verlag, 2001.

[154] S. Onn and M. Tennenholtz. Determination of social laws for multi-agent mobilization. *Artificial Intelligence*, 95(1):155–167, 1997.

[155] Organización Nacional de Transplantes. http://www.msc.es/ont.

[156] Orden de 2 de febrero de 1995 por la que se aprueba la primera relación de países con protección de datos de carácter personal equiparable a la española, a efectos de transferencia internacional de datos. Boletín Oficial del Estado 35, 10 de febrero 1995.

[157] Orden de 31 de julio de 1998 por la que se amplía la relación de países con protección de datos de carácter personal equiparable a la española, a efectos de transferencia internacional de datos. Boletín Oficial del Estado 106, 21 de agosto 1998.

[158] Organización Nacional de Transplantes, editor. *Informes y Documentos de Consenso promovidos por la Organización Nacional de Transplantes y la Comisión de Transplantes del Consejo Interterritorial del Sistema Nacional de Salud.* Editorial Complutense S. A., 1st edition, 2000.

[159] E. Ostrom. An agenda for the study of institutions. *Public Choice*, 48:3–25, 1986.

[160] O. Pacheco and J. Carmo. A role based model for the normative specification of organized collective agency and agents interaction. *Journal of Autonomous Agents and Multi-Agent Systems*, 6(2):145–184, march 2003.

[161] J. Padget. Modelling simple market structures in process algebras with locations. In Luc Moreau, editor, *AISB'01 Symposium on Software Mobility and Adaptive Behaviour*, pages 1–9. The Society for the Study of Artificial Intelligence and the Simulation of Behaviour, AISB, 2001. ISBN 1 902956 22 1.

[162] J. Padget. Modelling simple market structures in process algebras with locations. *Artificial Intelligence and Simulation of Behaviour Journal*, 1(1):87–108, 2001. ISSN 1476-3036.

[163] P. Panzarasa, N.R. Jennings, and T.J. Norman. Social mental shaping: Modelling the impact of sociality on autonomous agents' mental states. *Computational Intelligence*, 17(4), 2001.

[164] D. Jiménez Pastor. *HARMONIA: Automatic Generation of E-Organisations from Institution Specifications.* Department of Computer Science, Bath University, 2003. M.S. Thesis.

[165] D. Jiménez Pastor and J. Padget. Towards harmonia: automatic generation of e-organisations from institution specifications. In *Proceedings of Ontologies in Agent Systems at AAMAS03*, 2003. Published electronically at http://CEUR-WS.org/Vol-73/oas03-jimenez.pdf.

[166] E. Pattison-Gordon, J.J. Cimino, G. Hripsak, S.W. Tu, J.H. Gennari, N. L. Jain, and R. A. Greenes. Requirements of a sharable guideline representation for computer applications. Technical Report SMI-96-0628, Stanford Medical Institute, 1996.

[167] C. Penco. Objective and cognitive context. In *Modelling and Using Context: Proceedings of the Second International and Interdisciplinary Conference, CONTEXT'99, Trento, Italy*, Lecture Notes on Artificial Intelligence 1688, pages 270–283. Springer-Verlag, September 1999.

[168] M.E. Pollack. Mutual intention. *Artificial Intelligence*, (57):43–68, 1992.

[169] H. Prakken. *Logical Tools For Modelling Legal Argument. A Study of Defeasible Reasoning in Law.* Number 32 in Law and Philosophy Library. Kluwer Academic Publishers, 1997.

[170] H. Prakken. Modelling Defeasibility in Law: Logic or Procedure? *Fundamenta Informaticae*, (48(2-3)):253–271, 2001.

[171] H. Prakken. Modelling reasoning about evidence in legal procedure. In *Proceedings of the Eigths International Conference on Artificial Intelligence and Law, ICAIL 2001*, pages 119–128. ACM Press, May 21-25 2001.

[172] H. Prakken and G. Sartor. The role of logic in computational models of legal argument: A critical survey. In *Computational Logic: Logic Programming and Beyond*, Lecture Notes on Artificial Intelligence 2408, pages 342–381. Springer-Verlag, September 2002.

[173] M.J. Prietula, K.M. Carley, and L. Gasser, editors. *Simulating Organizations. Computational Models of Institutions and Groups*. AAAI Press/MIT Press, 1st edition, 1998.

[174] J.M. Pujol, R. Sangüesa, and J. Delgado. Extracting reputation in multi agent system by means of social network topology. *Proceedings of the First International Joint Conference on Autonomous Agents and Multi-Agent Systems AAMAS-02. Bologna, Italy*, 2002.

[175] V. Ram. *An E-Market System for Electric Utilities*. Department of Computer Science and Engineering, Arizona State University. M.S. Thesis.

[176] A.S. Rao and M.P. Georgeff. Modeling rational agents within a BDI-architecture. In James Allen, Richard Fikes, and Erik Sandewall, editors, *Proceedings of the 2nd International Conference on Principles of Knowledge Representation and Reasoning (KR'91)*, pages 473–484. Morgan Kaufmann publishers Inc.: San Mateo, CA, USA, 1991.

[177] A.S. Rao and M.P. Georgeff. An abstract architecture for rational agents. In *Proceedings of the 3rd International Conference on Principles of Knowledge Representation and Reasoning (KR'92)*, pages 439–449. Morgan Kaufmann publishers Inc.: San Mateo, CA, USA, 1992.

[178] A.S. Rao and M.P. Georgeff. BDI-agents: from theory to practice. In *Proceedings of the First International Conference on Multiagent Systems*, San Francisco, 1995.

[179] Real Decreto 1332/1994, de 20 de junio, por el que se desarrollan determinados aspectos de la ley orgánica 5/1992, de 29 de octubre, de Regulación del Tratamiento Automatizado de los Datos de carácter personal. Boletín Oficial del Estado 147, 21 de junio 1994.

[180] Real Decreto 2070/1999, de 30 de diciembre, por el que se regulan las actividades de obtención y utilización clínica de órganos humanos y la coordinación territorial en materia de donación y transplante de órganos y tejidos. Boletín Oficial del Estado 3, 4 de Enero 2000.

[181] Real Decreto-Ley 14/1999, de 17 de septiembre, sobre firma eletrónica. Boletín Oficial del Estado 224, 18 de septiembre 1999.

[182] Real Decreto 994/1999, de 11 de junio, por el que se aprueba el reglamento de medidas de seguridad de los ficheros automatizados que contengan datos de carácter personal. Boletín Oficial del Estado 151, 26 de febrero 2000.

[183] Recommendation No. R(97)5 of the Committee of Ministers to Member States on the Protection of Medical Data. Council of Europe, 13 February 1997.

[184] A. Rector, S. Bechhofer, C.A. Goble, I. Horrocks, W.A. Nolan, and W.D. Solomon. The GRAIL concept modelling language for medical terminology. *Artificial Intelligence in Medicine*, 9:139–171, 1997.

[185] Regulation (EC) No. 45/2001 of the European Parliament and of the Council of 18 december 2000 on the protection of individuals with regard to the processing of personal data by the Community institutions and bodies and on the free movement of such data. Official Journal of the European Communities, January 12 2001.

[186] REgional and International Integrated Telemedicine Network for Medical Assistance in End Stage Diseases and Organ TRANSPLANT. http://retransplant.vitamib.com.

[187] C. Rich and C.L. Sidner. COLLAGEN: When agents collaborate with people. In W. Lewis Johnson and Barbara Hayes-Roth, editors, *Proceedings of the First International Conference on Autonomous Agents (Agents'97)*, pages 284–291, New York, 5–8, 1997. ACM Press.

[188] J.A. Rodríguez-Aguilar. *On the Design and Construction of Agent-mediated Electronic Institutions*. PhD thesis, Institut d'Investigació en Intel.ligència Artificial (IIIA), 2001.

[189] J. Sabater and C. Sierra. Reputation and social network analysis in multi-agent systems. *Proceedings of the First International Joint Conference on Autonomous Agents and Multi-Agent Systems AAMAS-02. Bologna, Italy*, 2002.

[190] F.A.A. Santos and J. Carmo. *Indirect Action. Influence and Responsibility.* Springer, 1994.

[191] F.A.A. Santos, A.J.I. Jones, and J. Carmo. Action Concepts for Describing organised Interaction. In *Proceedings of the Thirtieth Annual Hawaii International Conference on System Sciences*. IEEE Computer Society Press, 1997.

[192] Scandiatransplant. http://www.scandiatransplant.org.

[193] M. Schillo, I. Zinnikus, and K. Fischer. Towards a Theory of Flexible Holons: Modelling Institutions for Making Multi-Agent Systems Robust. In R. Conte and C. Dellarocas, editors, *2nd Workshop on Norms and Institutions in Multi-Agent Systems (Agents 2001)*. ACM Press, 2001.

[194] A.M. Di Sciullo. Formal context and morphological analysis. In *Modelling and Using Context: Proceedings of the Second International and Interdisciplinary Conference, CONTEXT'99, Trento, Italy*, Lecture Notes on Artificial Intelligence 1688, pages 105–118. Springer-Verlag, September 1999.

[195] W.R. Scott. *Organizations: Rational, Natural and Open Systems.* Prentice Hall, 1998.

[196] W.R. Scott. *Institutions and Organizations.* Sage, 2nd edition, 2001.

[197] V. Shankararaman, V. Ambrosiadou, T. Panchal, and B. Robinson. Agents in health care. In V. Shankararaman, editor, *Workshop on Autonomous Agents in Health Care,* pages 1–11, 2000.

[198] Y. Shoham. Agent-Oriented Programming. *Artificial Intelligence,* 60(1):51–92, 1993.

[199] Y. Shoham and M. Tennenholtz. On social laws for artificial agent societies: Offline design. *Artificial Intelligence,* 73(1-2):231–252, 1995.

[200] E.H. Shortliffe. *Computer-Based Medical Consultations: MYCIN.* Elsevier, 1976.

[201] B.G. Silverman, C. Andonyadis, and A. Morales. Web-based health care agents; the case of reminders and todos, too (r2do2). *Artificial Intelligence in Medicine,* 14(3):295–316, 1998.

[202] J.M. Spivey. *The Z Notation: A Reference Manual.* Prentice Hall, London, 1992.

[203] M. Tambe. Towards flexible teamwork. *Journal of Artificial Intelligence Research,* 7:83–124, 1997.

[204] TECN : Transplant Euro Computer Network. http://www.ejeisa.com/nectar/t-book/html/health.htm#TECN.

[205] R.H. Thomason. Type theoretic foundations for context, part 1: Contexts as complex type-theoretic objects. In *Modelling and Using Context: Proceedings of the Second International and Interdisciplinary Conference, CONTEXT'99, Trento, Italy,* Lecture Notes on Artificial Intelligence 1688, pages 351–360. Springer-Verlag, September 1999.

[206] R. Tuomela. *The importance of Us: A Phylosophical Study of Basic Social Norms.* Standford University Press, 1995.

[207] R. Tuomela and K. Miller. We Intentions. *Philosophical Studies,* (53):367–389, 1988.

[208] R.M. Turner. A model of explicit context representation and use for intelligent agents. In *Modelling and Using Context: Proceedings of the Second International and Interdisciplinary Conference, CONTEXT'99, Trento, Italy,* Lecture Notes on Artificial Intelligence 1688, pages 375–388. Springer-Verlag, September 1999.

[209] United Network for Organ Sharing. http://www.unos.org.

[210] A. Valls, A. Moreno, and D. Sánchez. A multi-criteria decision aid agent applied to the selection of the best receiver in a transplant. In *4th. International Conference on Enterprise Information Systems, Ciudad Real, Spain.,* 2002.

[211] T.M. van Engers and G.C. van der Veer. POWER: an illustration of task based design for groupware. In R. Traunmüller and A. Rizzo, editors, *Workshop on*

Distributed Cognition and Distributed Knowledge: Key issues in design for e-Commerce and e-Government, 2000.

[212] W. Vasconcelos, J. Sabater, C. Sierra, and J. Querol. Skeleton-based agent development for electronic institutions. In *Proceedings of the First International Conference on Autonomous Agents and Multiagent Systems (AAMAS-02)*.

[213] J. Vázquez-Salceda. *The Role of Norms and Electronic Institutions in Multi-Agent Systems applied to complex domains. The HARMONIA framework*. PhD thesis, Artificial Intelligence PhD Program, Universitat Politècnica de Catalunya, Barcelona, Spain, 2003.

[214] J. Vázquez-Salceda, U. Cortés, and J. Padget. Formalizing an electronic institution for the distribution of human tissues. In *Artificial Intelligence in Medicine*, volume 23, pages 233–258. Elsevier, March 2003.

[215] J. Vázquez-Salceda, V. Dignum, and F. Dignum. Organizing multi-agent systems. Technical report, Institute of Information and Computing Sciences, Utrecht University, 2004.

[216] H.J.E. Verhagen. *Norm Autonomous Agents*. PhD thesis, Department of Computer and Systems Sciences, The Royal Institute of Technology and Stockholm University, Sweden, 2000.

[217] H.J.E. Verhagen and J. Kummeneje. Adjustable autonomy, delegation, and distribution of decision making. In *Proceedings of CEEMAS'99*, pages 301–306, 1999.

[218] O. Vickers and J. Padget. Skeletal JADE Components for the Construction of Institutions. In J. Padget, D. Parkes, N. Sadeh, O. Shehory, and W. Walsh, editors, *Agent Mediated Electronic Commerce IV*, volume 2531 of *Lecture Notes in Artificial Intelligence*, pages 174–192. Springer Verlag, December 2002.

[219] G.H. von Wright. Deontic logic. *Mind*, (60):1–15, 1951.

[220] R.T. Walton, C. Gierl, P. Yudkin, H. Mistry, M.P. Vessey, and J. Fox. Evaluation of computer support for prescribing (capsule) using simulated cases. *British Medical Journal*, (315(7111)):791–795, 1999.

[221] E. Werner. Cooperating agents: A unified theory of communication and social structure. *Distributed Artificial Intelligence Vol II*, pages 3–36, 1989.

[222] E. Werner. Logical foundations of distributed artificial intelligence. *Foundations of Distributed Artificial Intelligence*, 1996.

[223] M. Wooldridge. *An Introduction to Multiagent Systems*. John Wiley & Sons, 2002.

[224] M. Wooldridge and P. Ciancarini. Agent-oriented software engineering: the state of the art. In P. Ciancarini and M. Wooldridge, editors, *Agent-Oriented Software Engineering*, volume 1957 of *LNAI*. Springer-Verlag, 2001.

[225] M. Wooldridge, M. Fisher, M.P. Huget, and S. Parsons. Model checking multiagent systems with mable. In C. Castelfranchi and W. L. Johnson, editors, *Proceedings of the First International Conference on Autonomous Agents and Multiagent Systems (AAMAS-02)*, volume 2, pages 952–959. ACM, ACM Press, July 2002.

[226] M. Wooldridge and N.R. Jennings. Intelligent agents: Theory and practice. *The Knowledge Engineering Review*, 10(2):115–152, 1995.

[227] M. Wooldridge, N.R. Jennings, and D. Kinny. The Gaia Methodology for Agent-Oriented Analysis and Design. *Autonomous Agents and Multi-Agent Systems*, 3(3):285–312, 2000.

[228] J. Yen, J. Yin, T.R. Ioerger, M.S. Miller, D. Xu, and R.A. Volz. CAST: Collaborative agents for simulating teamwork. In *Proceedings of the International Join Conference on Artificial Itelligence (IJCAI'01)*, pages 1135–1144, 2001.

[229] B. Yu and M.P. Singh. A social mechanism of reputation management in electronic communities. *Proceedings of Fourth International Workshop on Cooperative Information Agents*, pages 154–165, 2000.

[230] B. Yu and M.P. Singh. Search in referral networks. In G. Lindemann, D. Moldt, M. Paolucci, and B. Yu, editors, *Proceedings of the International Workshop on Regulated Agent-Based Social Systems: Theories and Applications (RASTA'02), Bologna, Italy*, volume 318 of *Mitteilung*, pages 213–229, Hamburg, 12 July 2002. Fachbereich Informatik, Universität Hamburg.

[231] G. Zacharias and P. Maes. Trust management through reputation mechanisms. *Applied Artificial Intelligence*, (14):881–907, 2000.

List of Acronyms

AAAI	American National Conference on AI
AAMAS	International Conference on Autonomous Agents and Multi-Agent Systems
ACIA	Catalan Association on Artificial Intelligence (*Associació Catalana d'Intel.ligència Artificial*)
ACL	Agent Communication Language
ACM	Association for Computing Machinery
AI	Artificial Intelligence
AISB	The Society for the Study of Artificial Intelligence and the Simulation of Behaviour
AMEC	Workshop on Agent-Mediated Electronic Commerce
AOP	Agent-Oriented Programming
AOSE	Agent-Oriented Software Engineering
APD	Spanish Data Protection Agency (*Agencia de Protección de Datos*)
BDI	Beliefs-Desires-Intentions
CAS	Contractual Agent Societies
CCIA	Catalan Conference on Artificial Intelligence (*Congrés Català d'Intel.ligència Artificial*)
CFP	Call For Proposals protocol
CIL	Contextual Intensional Logic
CMB	Context-Mediated Behavior
CSIC	Spanish National Research Council (*Consejo Superior de Investigaciones Científicas*)
CTL	Computational Temporal Logic
DAI	Distributed AI
DAML	DARPA Agent Markup Language
DARPA	Defense Advanced Research Projects Agency
DB	DataBase
DFOL	Distributed First Order Logic
EMR	Electronic Medical Reports
EU	European Union
FIPA	Foundation for Intelligent Physical Agents
FOL	First Order Logic
HLA	Human Leukocyte Antigens
HSCSP	Hospital de la Santa Creu i Sant Pau
HTC	Human Transplant Coordinator
IIIA	Artificial Intelligence Research Institute (*Institut d'Investigació en Intel.ligència Artificial, CSIC*)

IAS	Intelligent Agent Societies
IEEE	Institute for Electrical and Electronic Engineers, Inc.
IJCAI	International Joint Conference on AI
JADE	Java Agent DEvelopment framework
KB	Knowledge Base
LGS	Spanish Health Law (*Ley General de Sanidad*)
LMS	Local Model Semantics
LOPD	Spanish Organic Law on Data Protection (*Ley Orgánica de Protección de Datos*)
LORTAD	Former Spanish Organic Law on Data Protection (*Ley Orgánica de Regulación del Tratamiento Automatizado de los Datos de carácter personal*)
MAS	Multi-Agent Systems
MIT	Massachussetts Institute of Technology
NATO	North-Atlantic Treaty organization
NEI	american National Eye Institute
NLP	Natural Language Processing
NuSMV	New Symbolic Model Verifier
OCATT	Catalan Transplants organization (*Organització CATalana de Trasplantaments*)
ONT	Spanish National Transplants organization (*Organización Nacional de Transplantes*)
PDA	Personal Digital Assistant
PLC	Propositional Logic of Context
RASTA	International Workshop on Regulated Agent-based Social systems: Theories and Applications
RBAC	Role-Based Access Control
RETRANSPLANT	REgional and international integrated telemedicine network for medical assistance in end stage diseases and organ TRANSPLANT
SMS	Short Message Service
TECN	Transplant Euro Computer Network
UCT	an hospital's Transplant Coordination Unit (*Unidad de Coordinación de Transplantes*)
UML	Unified Modelling Language
UNOS	United Network for Organ Sharing
WAP	Wireless Application Protocol
WSES or WSEAS	World Scientific and Engineering Academy and Society

Glossary

Authors Index

Whitestein Series in Software Agent Technologies

Your Specialized Publisher in Mathematics

Birkhäuser

■ **Günter, M.**, Zürich, Switzerland

Customer-based IP Service Monitoring with Mobile Software Agents

2002. 168 pages. Softcover.
ISBN 3-7643-6917-5

Presenting mobile software agents for Internet service monitoring, this research monograph discusses newly standardized Internet technologies that allow service providers to offer secured Internet services with quality guarantees. Yet, today the customers of such services have no independent tool to verify (monitor) the service quality. This book shows why mobile software agents are best fit to fill the gap.
Key features:
- An introduction to standard Internet service enabling and managing technology such as IPSec, DiffServ and SNMP
- A generic service monitoring architecture based on mobile agents
- An object-oriented implementation of the architecture based on the Java programming language
- Several implementations of mobile software agents that can monitor new and emerging Internet services such as virtual private networks (VPN)

■ **Calisti, M.**, Zürich, Switzerland

An Agent-Based Approach for Coordinated Multi-Provider Service Provisioning

2002. 292 pages. Softcover.
ISBN 3-7643-6922-1

Communication networks are very complex and interdependent systems requiring complicated management and control operations under strict resource and time constraints. A finite number of network components with limited capacities need to be shared for dynamically allocating a high number of traffic demands. Moreover, coordination of peer provider is required whenever these demands span domains controlled by distinct operators. In this context, traditional human-driven management is becoming increasingly inadequate to cope with the growing heterogeneity of actors, services and technologies populating the current deregulated market.
This book proposes a novel approach to improve multi-provider interactions based on the coordination of autonomous and self-motivated software entities acting on behalf of distinct operators. Coordination is achieved by means of distributed constraint satisfaction techniques integrated within economic mechanisms, which enable automated negotiations to take place. This allows software agents to find efficient allocations of service demands spanning several networks without having to reveal strategic or confidential data. In addition, a novel way of addressing resource allocation and pricing in a compact framework is made possible by the use of powerful resource abstraction techniques.

■ **Moreno, A.**, Tarragona, Spain / **Nealon, J.L.**, Oxford, U.K. (eds.)

Applications of Software Agent Technology in the Health Care Domain

2003. 212 pages. Softcover.
ISBN 3-7643-2662-X

This volume contains a collection of papers that provides a unique, novel and up-to-date overview of how software agents technology is being applied in very diverse problems in health care, ranging from community care to management of organ transplants. It also provides an introductory survey that highlights the main issues to be taken into account when deploying agents in the health care area.